Networks in Action

International Series in Operations Research & Management Science

Volume 140

Series Editor:
Frederick S. Hillier
Stanford University, CA, USA

Gerard Sierksma • Diptesh Ghosh

Networks in Action

Text and Computer Exercises in Network Optimization

 Springer

Gerard Sierksma
Fac. of Economics and Business
University of Groningen
9700 AV Groningen
The Netherlands
g.sierksma@rug.nl

Diptesh Ghosh
Department of Production and
  Quantitative Methods
Indian Institute of Management
Ahmedabad-380015
Vastrapura, India
diptesh@iimahd.ernet.in

ISSN 0884-8289
ISBN 978-1-4614-2543-4   e-ISBN 978-1-4419-5513-5
DOI 10.1007/978-1-4419-5513-5
Springer New York Dordrecht Heidelberg London

Springer is part of Springer Science+Business Media (www.springer.com)

Preface

One of the most well-known of all network optimization problems is the shortest path problem, where a shortest connection between two locations in a road network is to be found. This problem is the basis of route planners in vehicles and on the Internet. Networks are very common structures; they consist primarily of a finite number of locations (points, nodes), together with a number of links (edges, arcs, connections) between the locations. Very often a certain number is attached to the links, expressing the distance or the cost between the end points of that connection.

Networks occur in an extremely wide range of applications, among them are:

road networks;	cable networks;
human relations networks;	project scheduling networks;
production networks;	distribution networks;
neural networks;	networks of atoms in molecules.

In all these cases there are "objects" and "relations" between the objects. A network optimization problem is actually nothing else than the problem of finding a subset of the objects and the relations, such that a certain optimization objective is satisfied.

Why a book with computer exercises on network optimization? First of all, network problems in practice are mostly very large and extremely complicated. Only with the assistance of computers high quality solutions can be obtained. Most books in this field only offer exercises that support the understanding of the theory. The case studies, that are discussed in the literature, are usually not suitable for classroom analysis, since the data sets are missing, or are much too large to handle in a classroom setting.

This book contains a wide range of not too large network optimization problems, that need to be analyzed and solved by using the computer. The emphasis of the book is not on solution techniques. The reader may find her favorite software on the Internet. Some suggestions for associated readings and Internet sites related to optimization are provided in Chapter C of the book.

Even the first theoretical chapter is written in the spirit of the book. The reader learns the basic concepts within the framework of networks and computer techniques.

The prerequisites for this book are minimal. All theoretical concepts are clearly explained. However, the reader is highly recommended to use this book in combination with a textbook on the subject. Our suggestion is *Graphs, Networks, and Algorithms* by D. Jungnickel [18]. This book contains a collection of major network optimization problems together with the main solution techniques. It also contains a wide range of interesting examples. The book is clearly written and very accessible. (Further suggestions for textbooks are given in the Chapter C of the book.) The literature list contains suggestions on this subject as well. We advice the reader to examine some introductory graph theory and linear programming/optimization theory.

What actually is the subject that this book refers to? An Internet search on the term "network optimization" does not result in many textbooks. The reason is obvious. Network optimization is related to a broad range of disciplines, including:

quantitative logistics;	supply chain logistics;
combinatorial optimization;	network theory;
integer programming;	integer linear programming;
operations research;	management science;
discrete mathematics;	finite mathematics;
algorithmic graph theory;	computer science.

The closest related disciplines to network optimization are combinatorial optimization and quantitative logistics. Combinatorial optimization deals with problems where solutions are combinations of the "objects" and the "relations" in the problem. The number of such solutions is finite but typically very large. A decision maker needs to choose a solution that satisfies a pre-specified objective. Quantitative logistics deals with logistics and supply chain management problems that have been formulated in mathematical terms. So literature on the subjects of combinatorial optimization and quantitative logistics are usually the best choices to know more about network optimization.

In order to avoid that the book would become a disconnected set of exercises, we use the fictitious company Global Telecom Company (GTC) as a common theme throughout. The international market for telecommunication is still one of the fastest growing markets in the world. There is an increasing demand for more and more customer-specific products and services. Also in this market, the supply chains from product-suppliers via manufacturers, distribution centers, warehouses and retailers to the final customers have made a shift from "make-to-stock" to the customer-specific "make-to-order" concept. Outsourcing and focusing on core-business have become the key success factors. Demand-supply chain management, where all links in the production network cooperate, is a key issue for surviving and for obtaining the necessary competitive advantage over other supply networks. Finding the balance between on one hand maximizing the customer service, and on the other hand minimizing the total costs, wastes, and inventories is done on a supply chain/network scale. GTC operates in this scenario.

GTC has businesses in various countries across the world. The company manages a large worldwide telecommunication system, and produces sophisticated equipment for telecommunication. GTC operations are divided into a number of largely independent operation divisions, such as "Research & Development", "Cables", and "Services". One of the main goals of GTC is to extend its position as an important supplier of telecommunication services, with an emphasis on a continuous improvement of the price-quality relations of its products and services.

These facts justify a look at network optimization from a telecommunication point of view. The problems have been carefully selected and formulated and reflect a high degree of realism, although the reality is not always reached completely. When faced with the choice between realism on one hand and didactic justification on the other, we have often chosen for the latter. We have formulated problems with data sets that are manageable and surveyable for the student, but too large to be solved "by inspection".

The exercises in the book have been extensively tested in classroom settings. Many of our students' suggestions to improve the clarity of the questions have been implemented. However, we are eager to obtain further suggestions on improving the book. These may be sent by email to the authors' addresses (`g.sierksma@rug.nl` and `diptesh@iimahd.ernet.in`). Nevertheless, some questions may not be completely clear to the reader. In such cases, the reader is challanged to formulate her/his own interpretation of the question. Even a number of alternative scenarios may be considered. In practical situations, when the management is not completely clear about the formulation of a problem, a pro-active approach of proposing alternative scenarios may be effective, especially in understanding the problem more clearly.

Many persons have contributed to the final version of this book. We would especially like to mention the contributions of Matthijs Streutker, who provided the first version of the chapter on facility location, and devised the software for solving the problems, and Marc-Jan Menkhorst for providing the first version of the chapter on matching. Any errors that remain in the book are of course the responsibility of the authors.

Groningen, Gerard Sierksma
Ahmedabad, Diptesh Ghosh

Contents

A

The Modeling and Implementing Process

Mathematical models do not exist in vacuum, but are typically representations of real-life problem situations. Therefore, even in a book of problems in network optimization, it is important to point out the position of optimization modeling and solution in real life problem solving. Hence this chapter.

Modeling and solving a practical problem is usually a long process of advancing insight. The person on the work floor, who actually faces the problem, often intuitively feels the presence of a problem, and gradually, in a step-by-step process, identifies the actual problem. Only then it is time to think of solving the problem, and possibly implementing a computer decision support system to help the organization more accurate and effective in the future when similar problems arise. Very often organizations do not take enough time to analyze the problem in enough detail and buy expensive software that does not meet their needs sufficiently.

Decision-making begins with a situation in which a certain problem is recognized. The modeling process starts by calling on an analyst (or a team of analysts). The analyst formulates the problem by constructing a precise verbal statement of objectives, constraints on solutions, appropriate assumptions, descriptions of processes, data requirements, alternatives for action, and metrics for measuring progress. This process of formulation is extremely important: it does not make sense to solve a problem that is not the problem of the company.

The next step is to translate the problem from verbal, qualitative terms into a logical, quantitative model. Such a model is a number of rules, usually embedded in a computer program. A mathematical model, for instance a network model, is a collection of functional relationships. Models usually never include every aspect of the problem to be solved. However, they should include the more important and relevant aspects. Including too few aspects often leads to very elegant models that do not solve the real problem, while including too many aspects may lead to a realistic model that is too complicated to solve in reasonable time. The process of restricting the model to the most relevant aspects also forces the decision maker to set priorities, and using the solutions as bench marks when the less important aspects are taken into account, if necessary. The statements of the abstractions, introduced in the construction of the model, are called the assumptions. One of the main questions always is:

G. Sierksma and D. Ghosh, *Networks in Action: Text and Computer Exercises in Network Optimization*, International Series in Operations Research & Management Science 140, DOI 10.1007/978-1-4419-5513-5_1, © Springer Science + Business Media, LLC 2010

Does the model represent the relevant aspects of the problem? Or, do we solve an abstraction too far away from reality?

It may be clear that the above process is only possible in good cooperation between the decision maker(s) and the analyst(s). Otherwise, the assumptions may be become an unnecessary obstacle between the analyst and the decision maker. The analyst may be only willing to use the computer solutions, if he/she agrees on the assumptions.

During the modeling process the analyst may calculate temporary solutions, just as to validate the modeling process. If a solution does not make sense at all to the decision maker, the modeling process needs to be reconsidered. As soon as the analyst and the decision maker agree, the actual solution process of the model can start. This also includes the calculation and analysis of different scenarios, where the usually uncertain data is subject to change. This, so called 'what if' analysis highly contributes to the level of acceptance of the new system. The final most important aspect is of course the level of acceptance and the actual use of the system by the company.

Even when computer systems are applied, the users should always keep in mind that the non-tangible and qualitative aspects of the problems should not be neglected and denied; the systems implemented are indeed decision support systems, with a lot of emphasis on the term 'support'. Decision support systems are devoted to aiding decision makers under varying circumstances: limited resources, conflicting goals, changing conditions, complex interpersonal dynamics, uncertainty in for instance demand, and unyielding deadlines. The goal of computer systems is to provide a framework for the decision-making process, and to provide optimal solutions with respect to given measures of merit.

Since companies operate in a dynamic world, problems and solutions of yesterday may not be relevant for to day anymore. Therefore it is important to the company to establish controls that recognize a changing situation and signal the need to update the decision support system.

Finally, it seems to be a common human trait to resist against changes. Therefore the most tedious part is to get the system and the solutions implemented in the organization. So the art of building models also includes the art to carry along the people in the organization that are likely to be affected by the innovation.

B

Network Theory

B.1 Graphs and Networks

Network optimization needs to use many terms and notions used in graph theory. In this chapter we seek to introduce most of the graph theory terms and notions used in the book. We also introduce some concepts used in the study of algorithms. To explain these concepts, we will make use of a fictitious map of one of the facilities of the fictitious Global Telecom Company (GTC). The map is shown in Figure B.1 and is drawn to scale.

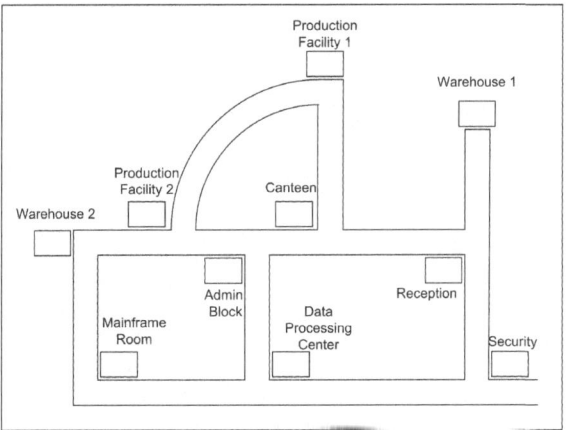

Fig. B.1. Map of one of the facilities at GTC (to scale)

If we want to depict the connectivity among various points in the facility, we can have a schematic representation of the map as shown in Figure B.2. In this figure, the locations of the map have been replaced by points, depicted as labeled circles. In Figure B.2 for example, point 1 represents the Security building in the map, point 2

G. Sierksma and D. Ghosh, *Networks in Action: Text and Computer Exercises in Network Optimization*, International Series in Operations Research & Management Science 140, DOI 10.1007/978-1-4419-5513-5_2, © Springer Science + Business Media, LLC 2010

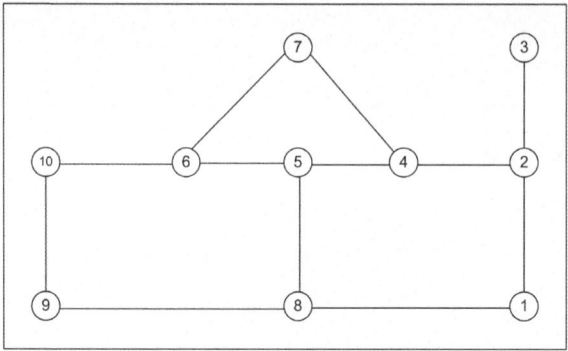

Fig. B.2. The map of Figure B.1 represented as a graph

represents the Reception desk, point 3 represents Warehouse 1, and so on. The lines between the points show that the buildings represented by the points are connected by a direct road segment. Figure B.2 is known as a *graph*. The points 1, 2, ..., 10 are referred to as *nodes* (or *vertices*) of the graph, and the lines between the nodes are referred to as *edges* (or *links*) in the graph. Hence a graph is just a collection of nodes and edges linking pairs of nodes. In some graphs, we allow more than one edges to connect a pair of nodes. As a special case of such graphs, an edge may connect a node to itself. Graphs with such possibilities are called multigraphs. Graphs that are not multigraphs are called *simple graphs*.

Notice that the representation that we made does not faithfully represent the geographical properties of the map; for example, the edges between 4 and 7, and 6 and 7 are drawn approximately equal, although the distance between the two production facilities in the map is distinctly more than the distance between Production Facility 1 and the Canteen. This is because the graph is just a representation that emphasizes more on the connectivity between various points rather than a faithful representation of geography.

If we want to convey information about the distances between each pair of points in the map, we make use of *weighted graphs* (see Figure B.3), also referred to as *networks*. Next to each edge in a weighted graph, there is a number denoting the weight (or cost, or length) of that edge. In Figure B.3, the weights denote the distances between the corresponding nodes in 100 meter units. So from it we can interpret that the distance between the Security building and the Reception desk is 200 meters.

Next let us assume that the road segments in Figure B.1 are one way roads. Let us suppose that one can go in the direction from the Security building to the Mainframe Room, but not in the other direction. The only other permissible directions of travel are from the Mainframe Room to Warehouse 2, down the straight road from Warehouse 2 to the Reception desk, from Production Facility 2 to Production Facility 1 to the Canteen, and from the Reception desk to Warehouse 1. These directional properties can be depicted in a *directed graph* (or *digraph*), which is a graph in which all edges have a direction assigned. Directed edges are also known as *arcs*. Figure B.4

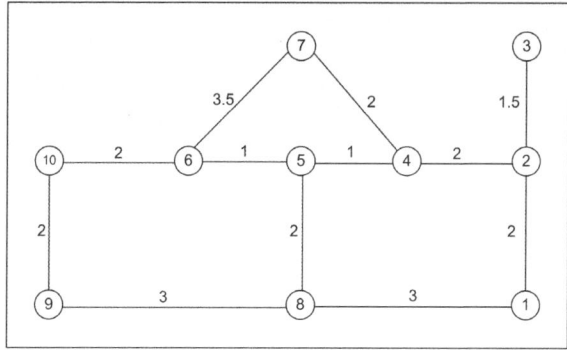

Fig. B.3. The map of Figure B.1 represented as a weighted graph

shows how the directed graph with all the directional properties mentioned above is depicted.

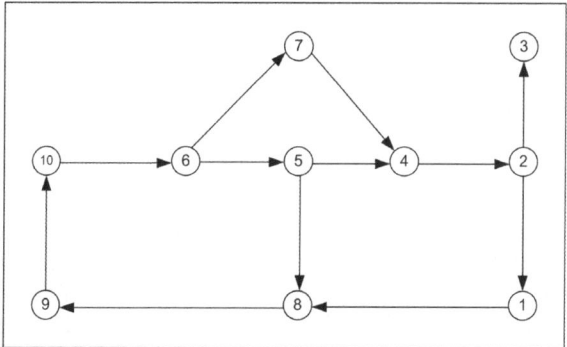

Fig. B.4. A directed graph representing the map of Figure B.1

Sometimes, in a graph, some (but not all) edges are directed. A graph with both directed and undirected edges is called a *mixed graph*. Obviously, undirected graphs and directed graphs are both special cases of mixed graphs. Mixed graphs, and hence directed graphs, have weighted versions too.

Simple graphs in which each pair of nodes is connected by an edge are called *complete graphs*. Therefore, a complete graph on n ($n \geq 2$) nodes (denoted by K_n) has $\binom{n}{2}$ edges. A complete graph on five nodes is shown in Figure B.5.

In some cases, it is possible to divide the the set of nodes in a graph into two disjoint sets, such that all edges in the graph join a node in one set to a node in the other, and no edge joins nodes in the same set. Such graphs are called *bipartite graphs*. Simple bipartite graphs, in which all nodes of one set are connected to each node of the other set, are called *complete bipartite graphs*. If one of the two sets of

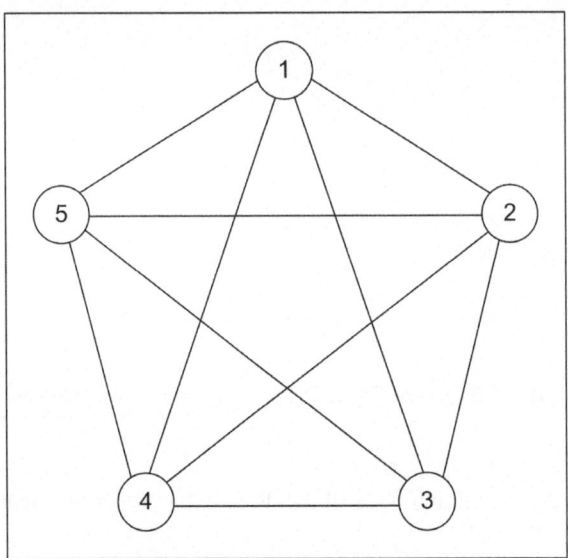

Fig. B.5. The complete graph K_5

nodes has m nodes in it, and the other set has n nodes, then the complete bipartite graph is denoted by $K_{m,n}$, $(m, n \geq 1)$. Figure B.6 depicts the complete bipartite graph $K_{3,2}$.

Let us consider the problem of traveling between nodes 5 and 9 in the graph of Figure B.2. There are several ways to do it. One may take a direct route from node 5 through nodes 6 and 10 to node 9. Another way could be from node 5 through nodes 6, 7, 4, and 10 to node 9. In graph theory, these routes are called walks. A *walk* is an alternating sequence of nodes and edges, starting and ending with nodes, where each edge joins the node preceding it with the node succeeding it. In this text we will denote the two walks as $5 - 6 - 10 - 9$, and $5 - 6 - 7 - 4 - 5 - 10 - 9$, respectively. A walk in which no two nodes are the same is called a *path*. So the walk $5 - 6 - 10 - 9$ is a path, while the walk $5 - 6 - 7 - 4 - 5 - 10 - 9$ is not. Walks and paths have equivalents in directed and mixed graphs (directed walk and mixed walk, and directed path and mixed path, respectively). A path in a graph which starts and ends at the same node is called a *cycle* or *tour*. In Figure B.2, $5 - 6 - 7 - 4 - 5$ denotes a cycle. A graph which does not contain a cycle is called an *acyclic graph*. The *length* of a walk, path, or cycle, is the sum of the weights of the edges in it for weighted graphs, and the number of edges in it for unweighted graphs. As in other cases, cycles have equivalents, called directed and mixed cycles, respectively, in directed and mixed graphs. Walks, paths, cycles, and tours can also be described by the edges or arcs in them. For example, the path $5 - 6 - 10 - 9$ can be denoted by the set $\{5 - 6, 6 - 10, 10 - 9\}$.

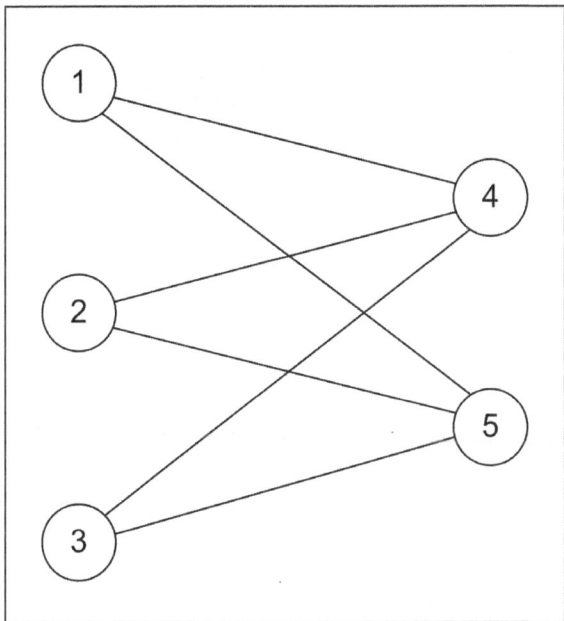

Fig. B.6. The complete bipartite graph $K_{3,2}$

B.2 Solution Techniques

In network optimization (and indeed in many other fields) we need to define systematic procedures to do particular operations. In most cases, we need the procedure to be finite, i.e., if one follows the procedure, one should be able to finish the task in finite time. Such well defined procedures that get over in finite time are called algorithms. They are called *exact* if they are guaranteed to output optimal solutions, otherwise they are called *approximate*. Approximate algorithms are also called *heuristics*.

Now, let us consider the problem of finding a shortest path between two pre-specified nodes in a graph. For example, let us suppose that one wants to find a shortest path between Warehouse 1 and Warehouse 2 in the map in Figure B.1. If one is willing to look at all paths, compare them and output the cheapest, then one has to consider the following seven options (in terms of nodes in the graph in Figure B.2):

(1) $3 - 2 - 1 - 8 - 9 - 10$;
(2) $3 - 2 - 1 - 8 - 5 - 6 - 10$;
(3) $3 - 2 - 1 - 8 - 5 - 4 - 7 - 6 - 10$;
(4) $3 - 2 - 4 - 7 - 6 - 10$;
(5) $3 - 2 - 4 - 7 - 6 - 5 - 8 - 9 - 10$;
(6) $3 - 2 - 4 - 5 - 6 - 10$; and
(7) $3 - 2 - 4 - 5 - 8 - 9 - 10$.

If we generalize this problem to a weighted complete graph with n nodes ($n \geq 2$), we see that there are $\sum_{i=0}^{n-2} \frac{(n-2)!}{(n-2-i)!}$ different paths between any two nodes. (It is left to the reader to check this formula.) This is a daunting number — even if we employed a computer that could enumerate and evaluate a million paths each second, it would take 551.86 years to find a shortest path in a weighted K_{20} graph.

Algorithms that evaluate all solutions and return the best are called *complete enumeration* (or *exhaustive enumeration*) algorithms, and are normally the algorithms of last resort. Typically, there are much smarter ways of solving these problems. For example, there is an algorithm for the shortest path problem due to E.W. Dijkstra (which we describe in Chapter 1) which performs an order of n^2 computations. In Table B.1 we illustrate the comparison between $\sum_{i=0}^{n-2} \frac{(n-2)!}{(n-2-i)!}$ and n^2 as n increases.

Table B.1. The growth of the functions $\sum_{i=0}^{n-2} \frac{(n-2)!}{(n-2-i)!}$ and n^2 as n increases

n	$\sum_{i=0}^{n-2} \frac{(n-2)!}{(n-2-i)!}$	n^2
2	1	4
3	2	9
4	5	16
6	65	36
8	1957	64
10	109601	100
20	1.74×10^{16}	400
30	8.29×10^{29}	900
40	1.42×10^{45}	1600
50	3.37×10^{61}	2500

Several network optimization problems, such as the shortest path problem, the minimum spanning tree problem, and the matching problem have such efficient algorithms. These problems are thus called "easy" problems, which effectively means that given the number of bits required to code the input data in binary digits (say k), an algorithm exists which will solve any instance of the problem input in a number of *elementary steps* which is a polynomial in k. For our purpose, elementary steps include adding or subtracting or multiplying or dividing or comparing two numbers (i.e., for any two numbers a and b, $a+b$, $a-b$, $a \times b$, $a \div b$, and " is $a \geq b$").

Unfortunately, there are network optimization problems for which it is not clear whether such algorithms exist. These are called "hard" (or NP-hard) problems. Examples of these problems are network location problems, like the k-median problem and the facility location problem (see Chapter 5), and some network routing problems, like the traveling salesperson problem (see Chapter 6). For hard problems, typically, the solution technique is a refinement of a complete enumeration algorithm. Consequently, in general, we are able to solve easy problems of sizes much larger than that of hard problems. For solving large instances of hard problems, one relies on heuristics, which aim to yield good quality solutions within reasonable times. Unlike exact algorithms that always output optimal solutions, and thus can be compared only on the basis of their execution time and computer memory usage, heuristics can also be compared on the basis of the quality of solutions they output.

B.3 Graph Representations

Graphs are typically stored in computers in one of two forms, an *adjacency matrix*, or an *incidence matrix*.

For a mixed graph with a node set $N = \{1,2,\ldots,n\}$ without weights, the adjacency matrix is a $n \times n$ matrix. If (and only if) the graph has an arc from node i to node j, then the adjacency matrix has a 1 in the (i,j) position, otherwise the position contains a 0. If the graph has an edge between two nodes i and j, then that edge is considered to be a shorthand for two arcs, one from i to j and another from j to i. Accordingly, the adjacency matrix has a 1 in the (i,j) position as well as the (j,i) position. As an example consider the graph shown in Figure B.7.

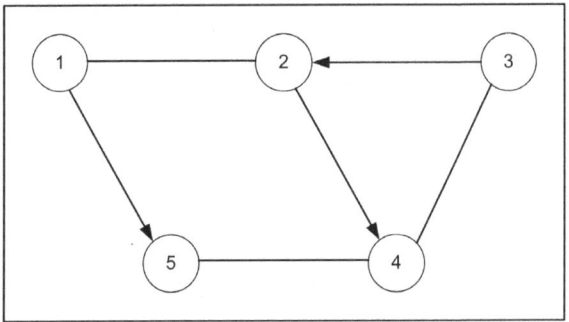

Fig. B.7. A simple graph

Its adjacency matrix representation is the following.

	1	2	3	4	5
1	0	1	0	0	1
2	1	0	0	1	0
3	0	1	0	1	0
4	0	0	1	0	1
5	0	0	0	1	0

If the graph is a weighted graph, then instead of a position in the matrix containing either a 1 or a 0, the position would either contain a 0 for a nonexistent arc, or the weight of that edge.

The maximum number of edges in a graph with n nodes is $n(n-1)$. If the graph has far fewer than this number of edges, then the representation can be made more efficient by using incidence matrices. For a graph with n nodes and m edges, an incidence matrix is a $n \times m$ matrix, where each row corresponds to a node and each column to an edge. In the column (say column k) corresponding to an edge connecting nodes i and j, only the (i,k) and (j,k) positions have a 1, and the others have a 0. The incidence matrix representation for the graph in Figure B.2, for example, is given below. Observe that this representation is *less* efficient than the adjacency matrix representation if the number of edges in the graph exceeds the number of nodes.

	1–2	1–8	2–3	2–4	4–5	4–7	5–6	5–8	6–7	6–10	8–9	9–10
1	1	1	0	0	0	0	0	0	0	0	0	0
2	1	0	1	1	0	0	0	0	0	0	0	0
3	0	0	1	0	0	0	0	0	0	0	0	0
4	0	0	0	1	1	1	0	0	0	0	0	0
5	0	0	0	0	1	0	1	1	0	0	0	0
6	0	0	0	0	0	0	1	0	1	1	0	0
7	0	0	0	0	0	1	0	0	1	0	0	0
8	0	1	0	0	0	0	0	1	0	0	1	0
9	0	0	0	0	0	0	0	0	0	0	1	1
10	0	0	0	0	0	0	0	0	0	1	0	1

If the graph is a directed graph, then the incidence matrix is called a *node arc incidence matrix*. The node arc incidence matrix of a directed graph with n nodes and m arcs has n rows and m columns. If the kth column represents an arc from i to j, then the (i,k) position of the matrix contains a -1 (instead of a 1 as in the earlier case) and the (j,k) position contains a 1 as in the previous case. Thus a node arc incidence matrix is a (-1, 0, 1)-matrix while the incidence matrix is a (0, 1)-matrix. The node arc incidence matrix representation for the graph in Figure B.4, for example, is the following.

	2→1	1→8	2→3	4→2	5→4	7→4	6→5	5→8	6→7	10→6	8→9	9→10
1	1	-1	0	0	0	0	0	0	0	0	0	0
2	-1	0	-1	1	0	0	0	0	0	0	0	0
3	0	0	1	0	0	0	0	0	0	0	0	0
4	0	0	0	-1	1	1	0	0	0	0	0	0
5	0	0	0	0	-1	0	1	-1	0	0	0	0
6	0	0	0	0	0	0	-1	0	-1	1	0	0
7	0	0	0	0	0	-1	0	0	1	0	0	0
8	0	1	0	0	0	0	0	1	0	0	-1	0
9	0	0	0	0	0	0	0	0	0	0	1	-1
10	0	0	0	0	0	0	0	0	0	-1	0	1

Note that each column in this matrix has exactly two non-zero elements, a 1 denoting the *head* of the corresponding arc, and a -1 denoting the *tail* of the arc.

C

References with Comments

Apart from Jungnickel's book [18] mentioned in the preface, we suggest (in alphabetical order of the first author's surname) the following textbooks:

1. Linear Programming by V. Chvátal [11]
   This is one of the best introductions on linear programming and optimization. Most of the concepts and techniques are introduced using illustrative examples.
2. Introduction to Operations Research by F.J. Hillier and G.J. Lieberman [16]
   This is a standard work on general operations research.
3. Operations Research; Models and Methods by P.A. Jensen and J.F. Bard [17]
   This is again a standard reference on general operations research.
4. Linear Optimization and Extensions by J.M. Padberg [23]
   This book on linear optimization grew out of a series of lectures and contains a chapter that relates combinatorial optimization with linear optimization.
5. Theory of Linear and Integer Programming by A. Schrijver [25]
   This book provides a thorough mathematical treatment of the theory of linear and integer optimization. It is aimed at an advanced level.
6. Linear and Integer Programming; Theory and Practice by G. Sierksma [27]
   This book on linear programming/optimization contains a large number of case studies, including several on network optimization.
7. Operations Research: an Introduction by H.A. Taha [30]
   This is another standard work on general operations research.
8. Model Building in Mathematical Programming, and Model Solving in Mathematical Programming by H.P. Williams [31, 32]
   These two books on building and solving mathematical models are very well written, and cover aspects of model building and solving that are dealt with very cursorily in other introductory books. The first book also contains a number of case studies on mathematical modeling.
9. Operations Research: Applications and Algorithms by W.L. Winston [34]
   This again is a major standard work on general operations research.

G. Sierksma and D. Ghosh, *Networks in Action: Text and Computer Exercises in Network Optimization*, International Series in Operations Research & Management Science 140, DOI 10.1007/978-1-4419-5513-5_3, © Springer Science + Business Media, LLC 2010

Other more advanced books and papers on network optimization are the following (again in alphabetical order of the first author's surname). One of the following symbols follow each of the titles.

CO: meaning that this is a reference on combinatorial optimization;
GL: meaning that this is a reference on general logistics;
GN: meaning that this is a reference on the theory of graphs and networks;
IP: meaning that this is a reference on general integer programming;
OR: meaning that this is a reference on general operations research.

All references also have a short description of the main characteristics.

1. Local Search in Combinatorial Optimization edited by E. Aarts and J.K. Lenstra [1]. (CO)
   This book deals with local search techniques for finding good quality solutions for optimization problems. Practical instances of hard problems like the facility location problem and the traveling salesman problem that we cover in our book can be solved using local search methods.
2. Network Flows; Theory, Algorithms and Applications by R.K. Ahuja, T.L. Magnanti, and J.B. Orlin [2]. (CO)
   This book is one of the most complete works on network optimization. It can be used as a thorough and advanced reference for the problems in our book.
3. Graphs and Applications; an Introductory Approach by J.M. Aldous and R.J. Wilson [3]. (GN)
   This book arose out of a British Open University course on graphs and networks. It is very suitable as an introduction on the concepts used in our book
4. Linear Programming and Network Flows by M.S. Bazaraa, J.J. Jarvis, and H.D. Sherali [4]. (IP)
   A considerable portion of this book is devoted to network optimization problems and algorithms for such problems. It has less coverage than Ahuja et al., but covers individual topics more elaborately.
5. Routing and Scheduling of Vehicles and Crews; the State of the Art by L.D. Bodin, B.L. Golden, A.A. Assad, and M.O. Ball [5]. (CO)
   Although this paper is not very recent, it provides an elaborate overview of vehicle routing problems and solution techniques. It is a good reference for our chapter Cyclic Routing.
6. Graph Theory with Applications by J.A. Bondy and U.S.R. Murty [6]. (GN)
   This is, in our opinion, still one of the best accounts on the theory of graphs and networks. It is very suitable as secondary reading to our book.
7. Team Formation: Matching Quality Supply and Quality Demand by B.H. Boon and G. Sierksma [7]. (CO)
   This paper describes the modeling process of designing optimal soccer and volleyball teams by solving a matching problem. It is a good reference for our chapter Matchings.
8. Logistical Management: the Integrated Supply Chain Process by D.J. Bowersox and D.J. Closs [8]. (GL)

This book focuses on business logistics, which includes all activities to move products and information to, from, and between the members of the supply chain. This textbook may serve as a thorough introduction to the fundamental logistics management issues.

9. The Logic of Logistics; Theory, Algorithms, and Applications for Logistics Management by J. Bramel and D. Simchi-Levi [9]. (GL)
   This book mainly deals with the mathematical aspects of logistics: vehicle routing, inventory, and warehouse location.

10. Combinatorial Optimization by W.J. Cook, W.H. Cunningham, W.R. Pulleyblank, and A. Schrijver [10]. (CO)
    This book is written by four leading scholars in the field of network optimization. The book starts on an elementary level and proceeds quickly to a more advanced level.

11. Network and Discrete Location; Models, Algorithms, and Applications by M.S. Daskin [12]. (CO)
    This text introduces the reader to the key classical location problems. Parts of the book can be used as supplementary reading for our chapter Facility Location.

12. Handbook of Combinatorial Optimization by D.-Z Du and P. Pardalos [13]. (CO)
    This series of three handbooks deals with algorithmic approaches for discrete and combinatorial problems, and brings together different aspects of these fields, with emphasis on recent developments.

13. Optimization Algorithms for Networks and Graphs by J.R. Evans and E. Minieka [14]. (CO)
    This book contains an elementary treatment of the main network optimization problems, including a nice chapter on network location theory. This last chapter is especially recommended.

14. Combinatorial Optimization; Theory and Algorithms by B. Korte and J. Vygen [19]. (CO)
    This is a well-written advanced graduate text, covering the major topics in combinatorial optimization. Applications are mentioned only occasionally.

15. The Traveling Salesman Problem; a Guided Tour of Combinatorial Optimization edited by E.L. Lawler, J.K. Lenstra, A.H.G. Rinnooy Kan, and D.B. Shmoys [20]. (CO)
    The major topics of combinatorial optimization are treated from the perspective of the Traveling Salesperson problem. This creative idea makes the book very readable and is highly recommended as secondary reading on our book.

16. A First Course in Combinatorial Optimization by J. Lee [21]. (CO)
    This book can be seen as an introduction in the mathematics of combinatorial optimization. It targets mainly on the graduate level.

17. Discrete Location Theory by P.B. Mirchandani and R.L. Francis [22]. (CO)
    This book consists of twelve papers written by experts in the field of discrete location theory. The book is suitable as further reading for our chapter Facility Location.

18. Combinatorial Optimization: Algorithms and Complexity by C.H. Papadimitriou and K. Steiglitz [24]. (CO)

This book is still seen as one of the most comprehensive texts on combinatorial optimization. Special attention is payed to the techniques for analyzing the complexity of algorithms.

19. Combinatorial Optimization: Polyhedra and Efficiency by A. Schrijver [26]. (CO)
    This state-of-the-art reference work consist of three volumes. Volume A: Paths, Flows, Matchings. Volume B: Matroids, Trees, Stable Sets. Volume C: Multiflows, Disjoint Paths, Hypergraphs. Volume A is particularly suited as a reference reading for our book.

20. Routing Helicopters for Crew Exchanges on Offshore Locations by G. Sierksma and G.A. Tijssen [28]. (CO)
    This case study describes the model building and model solving aspects of a real-life vehicle routing problem with split demands.

21. Designing and Managing the Supply Chain; Concepts, Strategies, and Case Studies by D. Simchi-Levi, P. Kaminski, and E. Simchy-Levi [29]. (GL)
    Supply Chain Management concerns the efficient integration of all suppliers, factories, warehouses, and stores/retailers within a chain/network, in order to minimize the total system costs, and to maximize the final customer's service, such as to acquire a competitive advantage over all other competing supply chains/networks. This introductory book discusses the basic topics of Supply Chain Management, among others inventory management, logistics network design, distribution systems, and decision support systems.

22. Graphs: an Introductory Approach by R.J. Wilson and J.J. Watkins [33]. (GN)
    This is a very good introduction on the theory of graphs and networks.

23. Integer Programming by L.A. Wolsey [35]. (IP)
    This is a textbook suitable for advanced undergraduate and masters level programs. It is aimed at users of integer programming techniques, who wish to understand why problems are sometimes difficult to solve, and how nevertheless good solutions can be obtained.

Apart from the books and papers listed above, certain websites are also useful resources for our book. We list them below. This list is not comprehensive, but provides a starting point for optimization sites on the Internet.

- Frequently asked questions on linear programming/optimization:
  http://www.unix.mcs.anl.gov/otc/Guide/faq/
  linear-programming-faq.html
- Tutorial on graph theory:
  http://www.cs.usask.ca/resources/tutorials/csconcepts/
  1999_8/tutorial/
- Tutorial on linear programming/optimization (due to H. Greenberg):
  http://carbon.cudenver.edu/~hgreenbe/courseware/
  LPshort/intro.html
- A page devoted to the traveling salesperson problem (includes the CONCORDE TSP Solver):
  http://www.tsp.gatech.edu/

- On optimization software in general (decision tree for optimization sofware):
  `http://plato.la.asu.edu/guide.html`
- The COIN-OR project has a collection of open-source solvers:
  `http://www.coin-or.org/`
- NEOS project (`http://www-neos.mcs.anl.gov/`) has a collection of solvers:
  `http://www-neos.mcs.anl.gov/neos/solvers/index.html`
- Free linear programming/optimization solvers are pointed to in the Linear Programming FAQ as well as in the INFORMS resources page. Informs Resource Collection (formerly known as Mike Trick's page):
  `http://www.informs.org/resources/`
- The Sci.op-Research Newsgroup deals with general operations research problems (available through GOOGLE® groups):
  `http://groups.google.com/group/sci.op-research?hl=en`

1

Shortest Paths

1.1 Introduction

The traffic in a particular city is controlled by seven groups of technicians in seven major junctions of the city. These seven junctions are labeled A, B, C, D, E, F, and G, and their interconnections along the road network in the city are schematically depicted by the network in Figure 1.1. The numbers next to the connections in the

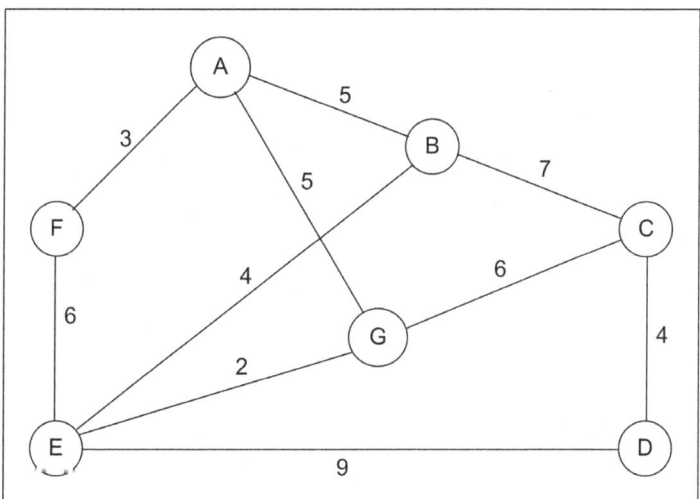

Fig. 1.1. The road network

network depict the time in minutes required to go from one end of the connection to the other end. The stations communicate with each other using sophisticated communication equipments. GTC holds a contract for the maintenance of this equipment. If there is a malfunction in the equipment at any of the junctions, GTC dispatches

G. Sierksma and D. Ghosh, *Networks in Action: Text and Computer Exercises in Network Optimization*, International Series in Operations Research & Management Science 140, DOI 10.1007/978-1-4419-5513-5_4, © Springer Science + Business Media, LLC 2010

maintenance crew from its base station at junction A to the junction at which the equipment malfunctions.

One major consideration for GTC is to minimize downtime, i.e., the time taken by GTC crew to reach the malfunctioning unit after they are informed of the malfunction. Therefore, the crew at location A need to know the quickest route through which they can reach any of the other six junctions if the need arises.

This problem is known as the *shortest path problem* in a network. It is one of the most common problems that occur in network optimization, and is very well studied. It is historically called the shortest path problem, since the cost associated with an edge in a network often represents the length of the edge.

Formally stated, in a shortest path problem, one is given a weighted network $N = (V, E, w)$ and two nodes $s, t \in V$ which are connected. w_e denotes the weight of the edge $e \in E$. These weights are often referred to as costs or lengths of the edges. A path $P = \{e_1, e_2, \ldots, e_k\}$ is said to have length $\sum_{e \in P} w_e$. The required output in the shortest path problem is a path P^* between s and t whose length is the minimum among the lengths of all paths between s and t in the network. If the network is a directed network, then one of s and t is designated as the *source*, and the other is designated as the *destination*. In directed networks the shortest path problem is one of finding a shortest length directed path from the source node to the destination node.

1.2 Applications

Several real-life problems can be modeled as shortest path problems. This section provides two such examples. Other situations where decisions are made using the shortest path problem are described in the problems at the end of the chapter.

1.2.1 Scheduling truck movement through cities

Trucks transporting bulk items on long inter-state routes are often either not allowed to pass through cities except during some restricted times of the day, or have to submit route plans that are approved by city authorities. In either case, transporters need to decide how to travel through the city so as to cause minimum disturbance. This problem can be modeled as a shortest path problem in which the roads that the trucks can take form a network, the entry point to the city is the source node, the exit point is the destination node, and the expected time for the truck to traverse a particular road segment is taken as its weight.

1.2.2 Making investment plans

Investors need to plan investments in order to maximize returns at a future date from a sum of money invested at present. Typically, they consider different investment options, each with its own rate of return and duration. The investor's problem can be solved using a variation of the shortest path problem. The modeling of the problem is not straightforward, hence the following example would be useful.

Consider that an investor is thinking of investing €1 and aims to maximize the returns from this investment at the end of three years. She has two investment options.

Option 1: Investment generates 10% (i.e., €0.10) returns with a duration of one year; and

Option 2: Investment generates 15% (i.e., €0.15) returns with a duration of two years.

She can invest only money that she has at hand at the beginning of each year. Notice that the total returns are not additive in the returns generated, but multiplicative. If the investor invests €1 in Option 1 at present and at the end of the first year takes the €1.10 it yields and invests in Option 2, then the total returns that she gets is *not* $[0.10+0.15]$, i.e. 25%, but $[(1.10) \times (1.15) - 1.00]$, i.e. 26.5%.

A possible way of modeling this is a network with four nodes, one labeled 0 representing the current time, and three others labeled 1, 2, and 3, representing the beginnings of the next three years. At each node i, $i \neq 3$, an arc is drawn to connect it to node $i+1$, representing the possibility of investing in Option 1 at the beginning of that year. In addition, at each node i, $i \neq 2, 3$, an arc is drawn to connect it to node $i+2$. These arcs represent the possibility of investing in Option 2 at the beginning of the year.

Since the return from an investment is computed by multiplying returns rather than adding them, the problem is converted to a shortest path problem by taking logarithms of the returns rather than the returns themselves. The length of a path in the network with these weights is then the logarithm of the return on a unit investment if the investor follows the investment policy represented by the path. The weight associated with each arc of the first type in the network is thus $\log(1.10)$ and the weight associated with each arc of the second type is $\log(1.15)$. The network for the investor's problem in the current setup is shown in Figure 1.2. The broken lines represent the arcs of the first type while the solid lines represent the arcs of the second type.

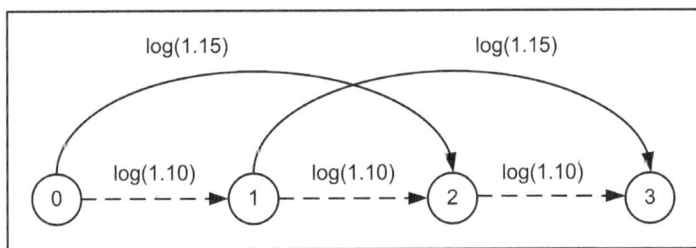

Fig. 1.2. Network denoting investment possibilities

A path on this network joining node 0 to node 3 describes an investment strategy; for example, the path $0 - 1 - 3$ denotes the strategy of investing in Option 1 now, and at the end of the first year, taking all the money and investing in Option 2. The sum of the weights of the arcs on each path is the logarithm of the return from the investment strategy represented by the path. The investor's objective is then to find the longest path in the network.[1]

1.3 A Linear Programming Formulation

Shortest path problems can be solved to optimality using the linear programming technique. The technique is best explained for directed networks. This does not pose a serious restriction to the applicability of linear programming to shortest path problems. An edge between two nodes i and j in a network, with a weight w_{ij}, can be represented as a pair of arcs, one from i to j and the other from j to i, both with a weight of w_{ij}. The linear programming problem then interprets the weight on each arc as the cost to transmit one unit of flow from the tail of the arc to the head, and tries to find a cheapest way of transmitting one unit of flow from the source node to the destination node. We denote the directed weighted network by $N = (V, A, w)$, the source node by s and the destination node by t.

In the linear programming formulation of the shortest path problem, a non-negative decision variable x_{ij} is defined for each arc $i \rightarrow j$ in the network. It represents the amount of flow through that arc if one unit of flow is to be sent as cheaply as possible from the source node to the destination node. Since the objective of the formulation is to obtain a cheapest route from s to t, it is represented as follows:

$$\text{Maximize} \sum_{i \rightarrow j \in A} w_{ij} x_{ij}. \tag{1.1}$$

The only set of constraints required in the formulation needs to ensure that for each node j in the network, the flow into node j equals the flow out node j, i.e.,

$$(\text{total inflow into node } j) = (\text{total outflow out of node } j).$$

These constraints are called *flow conservation constraints*. The flow into the source node and the flow out of the destination node are taken to be one unit each.

Conventionally, constraints are expressed with only a constant on the right hand side. Therefore the flow conservation constraints are modified in one of two different ways. For node j, either we can say that

$$(\text{total inflow into node } j) - (\text{total outflow out of node } j) = 0,$$

or we can say that

$$(\text{total outflow out of node } j) - (\text{total inflow into node } j) = 0.$$

---

[1] A similar model has also been used to decide on waste water treatment strategies in Kuwait. (See A.A. Elimam, D. Kohler, Two engineering applications of a constrained shortest-path model, European Journal of Operational Research, 103, (1997) pp. 426–438.)

Following the convention, the flow conservation constraint at node j in a network $N = (V, A, w)$ can be represented as follows:

$$\left. \begin{array}{l} \sum_{j \to k \in A} x_{jk} = 1 \text{ if } j = s \\ \sum_{i \to j \in A} x_{ij} = 1 \text{ if } j = t \\ \sum_{i \to j \in A} x_{ij} - \sum_{j \to k \in A} x_{jk} = 0 \text{ if } j \neq s, t \end{array} \right\} \tag{1.2}$$

The complete formulation to obtain a shortest path from node s to node t in a directed weighted network $N = (V, A, w)$ is shown in Figure 1.3. In this formulation we use flow conservation constraints in the form of constraint (1.2).

Minimize

$$z = \sum_{i \to j \in A} w_{ij} x_{ij}$$

Subject to

$$\sum_{s \to k \in A} x_{sk} = 1$$

$$\sum_{i \to t \in A} x_{it} = 1$$

$$\sum_{i \to j \in A} x_{ij} - \sum_{j \to k \in A} x_{jk} = 0 \text{ for each } j \neq s, t$$

$$x_{ij} \geq 0 \text{ for each } i \to j \in A$$

Fig. 1.3. Linear programming formulation of the shortest path problem

A shortest path from the source node s to the destination node t can be constructed from the optimal solution to this formulation. The optimal solution contains arcs $i \to j$ for which $x_{ij} > 0$, and other arcs $p \to q$ for which $x_{pq} = 0$. Any path from s to t using only such arcs $i \to j$ for which $x_{ij} > 0$ in an optimal solution to the linear programming formulation is a shortest path from s to t. If there is a unique shortest path, then all the decision variables in the optimal solution to the linear programming formulation have values of either 1 or 0. If there are multiple shortest paths, then some of the decision variables may take up non-integer values in an optimal solution to the linear program.

As an illustration of the formulation process, Figure 1.4 shows a linear programming formulation for obtaining a shortest path from location A to location D in the network in Figure 1.1. In each of the constraints, the inflows and outflows have been grouped using parentheses.

Minimize

$$z = 5(x_{AB} + x_{BA}) + 3(x_{AF} + x_{FA}) + 7(x_{BC} + x_{CB}) + 4(x_{CD} + x_{DC}) +$$
$$6(x_{CG} + x_{GC}) + 4(x_{BE} + x_{EB}) + 9(x_{DE} + x_{ED}) + 2(x_{EG} + x_{GE}) +$$
$$6(x_{EF} + x_{FE}) + 5(x_{AG} + x_{GA})$$

Subject to

$$x_{AB} - x_{AF} - x_{AG} = 1 \qquad \text{(Flow balance for A)}$$
$$x_{CD} + x_{ED} = 1 \qquad \text{(Flow balance for D)}$$
$$(x_{AB} + x_{CB} + x_{EB}) - (x_{BA} + x_{BC} + x_{BE}) = 0 \qquad \text{(Flow balance for B)}$$
$$(x_{BC} + x_{DC} + x_{GC}) - (x_{CB} + x_{CD} + x_{CG}) = 0 \qquad \text{(Flow balance for C)}$$

There are three more similar constraints for nodes E, F, and G.

$$x_{AB}, x_{BA}, x_{AF}, \dots, x_{FE}, x_{AG}, x_{GA} \geq 0 \qquad \text{(Non-negativity constraints)}$$

Fig. 1.4. Formulation of the shortest path problem for the road network in Figure 1.1

1.4 Algorithms for Shortest Path Problems

Linear programming implementations can solve shortest path problems, but the execution time required to solve such problems on large networks can be reduced significantly by implementing special purpose algorithms. Obviously, even though a very large number of paths exist between any two nodes in a moderate sized network, efficient algorithms look at only a very small subset of these paths to come up with a shortest path. In this section, two such algorithms are presented.

1.4.1 Dijkstra's algorithm

One of the most efficient algorithms for computing shortest paths is due to E.W. Dijkstra (1959). Consider a weighted directed network $N = (V, A, w)$, where the entries w_{ij} of w satisfy $w_{ij} \geq 0$ for any pair of nodes i and j in the network. The source and destination nodes in the network are labeled s and t, respectively. Dijkstra's algorithm works by iteratively finding a guaranteed shortest route from s to a node in the network for which the shortest path was not guaranteed in any previous iteration of the algorithm. It stops when the destination node t is the one to which it finds a shortest route at an iteration.

In Dijkstra's algorithm, each node in the network is associated with a label. The label associated with a node, say node j, has two parts. At any iteration of the algorithm, the first part stores the length of the shortest path found up to the end of the previous iteration of the algorithm from the source node s to node j. If no connection has been established until that point, then the length is taken to be ∞. Whenever the first part of the label at node j is finite, the second part of the label stores the predecessor of node j in the shortest path that the algorithm has found from s to j up to the end of the previous iteration. The predecessor of the source node s is taken to be s itself. In the following, the two parts of the label for node j are referred to as label_{j1} and label_{j2} respectively, and the label in its entirety is represented as a vector of the form $(\text{label}_{j1}, \text{label}_{j2})$.

At the beginning of any iteration in Dijkstra's algorithm, the nodes in V can be divided into two sets: V_1 such that for each $v \in V_1$, the algorithm has been able to guarantee that the path that it found from s to that node by the end of the previous iteration was indeed a shortest path to that node, and V_2 containing the remainder of the nodes. Notice that since all w_{ij} values are non-negative, the set V_1 always contains s, and is thus never empty.

During any iteration, the algorithm first considers each node v in V_2 to see if the shortest route that it has found to v from s could be improved in this iteration. To do this, it finds out whether there exists a node u in V_1, such that an arc $u \to v$ exists in A, and that $\text{label}_{v1} > \text{label}_{u1} + w_{uv}$. If no such node u exists in V_1, then the labels of the nodes in V_2 remain unchanged during the iteration. If one or more such nodes exist in V_1, then the algorithm chooses that u_0 in V_1 which minimizes the value $\text{label}_{u_01} + w_{u_0v}$. The value of label_{v1} is then updated by this value, and label_{v2} is replaced by u_0. Once the algorithm has finished trying to update the labels of all the nodes in V_2, it picks up a node, say v_0, in V_2 for which the value of label_{v_01} is the minimum among all nodes in V_2. It then transfers this node from V_2 to V_1. The reason for this transfer is the following. The shortest route from s to v_0 could either consist solely of intermediate nodes from V_1, or could include intermediate nodes from V_2. In the former case, the predecessor of v in such a shortest path would be a node from V_1, and since the algorithm has already guaranteed shortest paths to all nodes in V_1, therefore the shortest path to v is also guaranteed. If the shortest path contains any intermediate node v' in V_2, then the length of this path cannot be less than $\text{label}_{v'1}$. Since $\text{label}_{v_01} \leq \text{label}_{v'1}$, such a path cannot be shorter than the path already found by the algorithm. Hence the path to v found by the algorithm is guaranteed to be shortest. A pseudocode of Dijkstra's algorithm is given in Figure 1.5.

In order to illustrate the working of Dijkstra's algorithm, we refer to the network in Figure 1.1. The network is undirected, but can easily be converted into a directed network using the technique described at the beginning of Section 1.3. Consider the problem of finding a shortest path between A and D in the network. Initially, the label of node A is (0,A), and that of all other nodes is $(\infty, *)$. Set V_1 to $\{A\}$ and V_2 to $\{B, C, D, E, F, G\}$. In the first iteration, three arcs, namely $A \to B$, $A \to F$, and $A \to G$ are considered by the algorithm. Since label_{B1}, label_{F1}, and label_{G1} are all infinite, the labels at nodes B, F, and G are updated to (5,A), (3,A), and (5,A), respectively. Now, among the nodes in V_2, node F has the lowest value in the first part of its label. Hence F is transferred from V_2 to V_1. The second iteration thus starts with $V_1 = \{A, F\}$ and $V_2 = \{B, C, D, E, G\}$. The arcs considered at this iteration are $A \to B$, $A \to G$, and $F \to E$. Check that among these arcs, arc $A \to B$ and arc $A \to G$ cause no relabeling, while arc $F \to E$ causes the label of node E to change to (9,F). At this stage, among

Algorithm DIJKSTRA

Input: A directed weighted connected network $N = (V, A, w)$, a source node s and a destination node t.

Output: A shortest path in N from s to t.

Step 1 (Initialization):   Set the label of s to $(0, s)$ and the labels of all other nodes to $(\infty, *)$. Set $V_1 \leftarrow \{s\}$ and $V_2 \leftarrow V \setminus \{s\}$.

Step 2 (Termination):   If $t \in V_1$, then re-create the shortest path from s to t by starting from the second part of the label of t, and tracking predecessors along the shortest path, until s is reached. Output the shortest path and terminate. Else go to Step 3.

Step 3 (Iteration):   For each arc $u \rightarrow v \in A$ such that $u \in V_1$ and $v \in V_2$, check if $\text{label}_{v1} > \text{label}_{u1} + w_{uv}$. If so, then replace label_{v1} with $\text{label}_{u1} + w_{uv}$, and label_{v2} with u. Let $v_0 = \arg\min\{\text{label}_{v1} : v \in V_2\}$. Transfer node v_0 from V_2 to V_1. Go to Step 2.

Fig. 1.5. Pseudocode for Dijkstra's algorithm

all nodes in V_2, node B and G have the lowest value in the first part of their labels. At the end of the second iteration, node B is transferred from V_2 to V_1.

Figure 1.6 depicts the execution of Dijkstra's algorithm on the network in Figure 1.1. The label associated with each node is shown next to the node. To avoid cluttering the diagram, if a node has a label $(\infty, *)$, then the label is not shown in the diagram. Also, at the end of each iteration, the members of V_1 are shown using nodes colored gray, while the members of V_2 are shown using nodes colored white.

Dijkstra's algorithm belongs to a class of network algorithms called *label correcting algorithms*. At any point in the execution of this algorithm, each node carries a label that stores information about a shortest path found to that node thus far, and at each iteration of the execution, some of the labels could be updated depending on whether a shorter path to that node has been found.

Also notice that in the algorithm a shortest path to exactly one of the nodes of the network is obtained at the end of each stage. This node is the one that is transferred from V_2 to V_1 at the end of the iteration. For a network with n nodes, we are therefore sure that a shortest path between any two nodes in a network would be found in at most n iterations, and in each iteration no more alternatives than the number of arcs in the network would be checked. Thus, this algorithm is quite efficient in terms of execution time, even for large networks.

The shortest path problem is not well-defined for networks with edges with negative weights. If a network has an edge with negative weight, then the length of a shortest path between any two nodes of the network can be made arbitrarily small by traversing the negative weight edge a sufficient number of times. So Dijkstra's algorithm obviously fails on these types of networks.

However, the algorithm may also fail in a directed network with negative arcs. One of the reasons for its failure is the assumption in the initial step that the shortest path to the source node is guaranteed to be zero. This assumption is made when in Step 1 of Dijkstra's algorithm (see Figure 1.5) we set label_s to 0, and include s

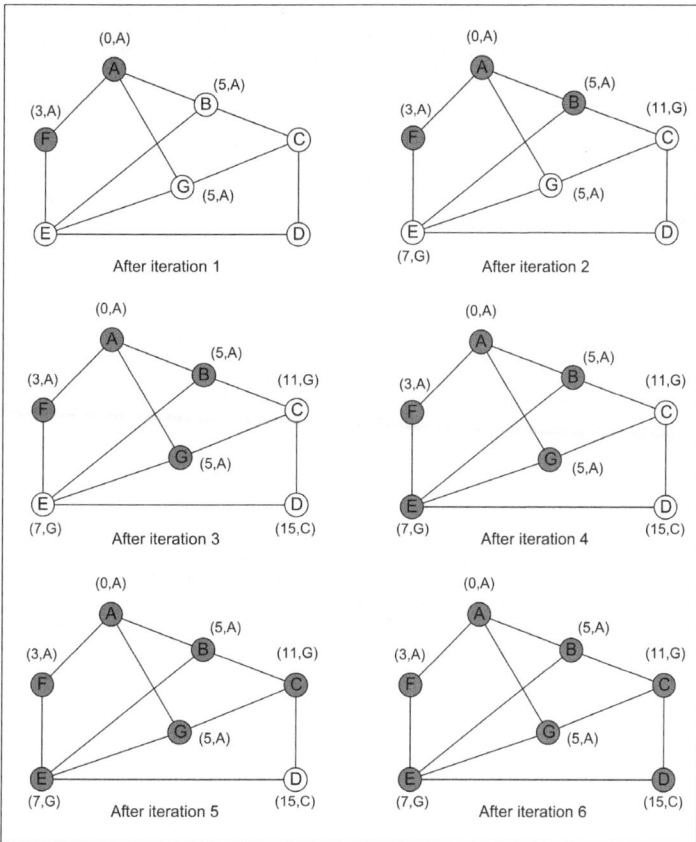

Fig. 1.6. Dijkstra's algorithm in action

in V_1. This assumption may not be valid if the network has a negative weight arc. Consider the network shown in Figure 1.7. If the source node for the shortest path in

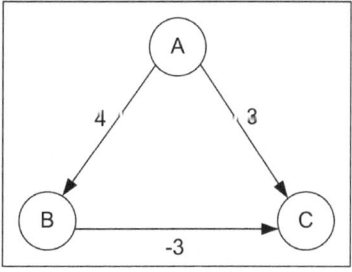

Fig. 1.7. A network on which Dijkstra's algorithm fails

the network is A and the destination node is C, then it is easy to see that Dijkstra's algorithm returns the path A – C with length 3, while a shorter path A – B – C with length 1 exists.

Thus there is a need for an algorithm to deal with directed networks with negative arc weights. Such an algorithm is described in the next subsection.

1.4.2 Bellman-Ford's algorithm

The algorithm due to R.E. Bellman (1958) and L.R. Ford (1962) uses labels similar to those used in Dijkstra's algorithm. However, it does not make use of the partitioning of nodes that is used in Dijkstra's algorithm. Bellman-Ford's algorithm starts by setting the label of the source node s to $(0,s)$ and the labels of all other nodes to $(\infty,*)$. During each iteration, it considers *all* arcs in the network. For an arc $u \to v$, if $label_{v1} > label_{u1} + w_{uv}$, then the label of node v is changed to $(label_{u1} + w_{uv},u)$. At the end of each iteration, the algorithm checks if the label of any of the nodes had changed during the iteration. If some label had changed, then the next iteration starts; otherwise, the first part of the label at any node u is the length of a shortest path to u from the source node s, and the second part of the label is the predecessor to u on a shortest path from s to u. The shortest path can then be traced backwards from u using the second part of the labels. A pseudocode of Bellman-Ford's algorithm is given in Figure 1.8.

Algorithm BELLMAN-FORD
Input: A directed weighted connected network $N = (V,A,w)$, a source node s and a destination node t.
Output: A shortest path in N from s to t.

Step 1 (Initialization):   Set the label of s to $(0,s)$ and the labels of all nodes except s to $(\infty,*)$. Set $iter\_ctr \leftarrow 0$ and $change\_flag \leftarrow$ FALSE.
Step 2 (Termination):   If $iter\_ctr = 0$, then go to Step 3. Else if $change\_flag \leftarrow$ FALSE, then re-create the shortest path from s to t by starting from the second part of the label of t, and tracking predecessors along the shortest path, until s is reached. Output the shortest path and terminate. Else set $change\_flag \leftarrow$ FALSE and go to Step 3.
Step 3 (Iteration):   For each arc $u \to v \in A$ check if $label_{v1} > label_{u1} + w_{uv}$. If so, then replace $label_{v1}$ with $label_{u1} + w_{uv}$, $label_{v2}$ with u, and set $change\_flag \leftarrow$ TRUE. Set $iter\_ctr \leftarrow iter\_ctr + 1$. Go to Step 2.

Fig. 1.8. Pseudocode for Bellman-Ford's algorithm

As an illustration of this algorithm, consider a directed network with negative arc weights as shown in Figure 1.9. Assume that the source node is A and the destination node is D. Also assume that the algorithm considers arcs in the order $A \to B, A \to F$, $B \to C, C \to D, C \to G, E \to B, E \to D, E \to G, F \to E, G \to A$.

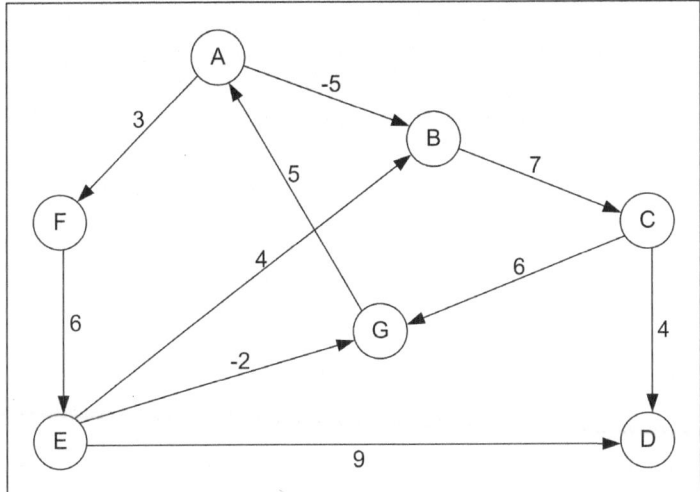

Fig. 1.9. A directed network with negative weight arcs

Initially the label of A is (0,A) and that of each of the other nodes is $(-\infty,*)$. In the first iteration, $A \rightarrow B$ causes the label of B to be set to $(-5,A)$, $A \rightarrow F$ causes the label of F to be set to (3,A), $B \rightarrow C$ causes the label of C to be set to (2,B), $C \rightarrow D$ causes the label of D to be set to (6,C), $C \rightarrow G$ causes the label of G to be set to (8,C), and $F \rightarrow E$ causes the label of E to be set to (9,F). None of the other arcs cause any relabeling at any of the nodes in the network. At the end of the second iteration, the labels of the nodes A through G are (0,A), $(-5,A)$, (2,B), (6,C), (9,F), (3,A), and (7,E); the third iteration does not cause any change in the labels. Therefore at the end of the third iteration, the algorithm terminates and outputs the shortest path from A to D as A→B→C→D with length six units.

Notice that at the end of the execution of this algorithm, the shortest paths from A to all other nodes in the network are also obtained. In this regard, the Bellman-Ford algorithm is more general than Dijkstra's algorithm. On the other hand, since at each iteration, the Bellman-Ford checks all the arcs in the network for possible relabeling opportunities, it is slower in execution than Dijkstra's algorithm.

1.5 Other Path Problems

1.5.1 The longest path problem

In this chapter, we have primarily dealt with the problem of finding a shortest path between any pair of nodes in a network. However sometimes, finding the longest path between a pair of nodes in a network becomes important.

Consider for example, a network describing the interconnections between tasks in a complex project. Each arc in the network represents a task, and the weight of an arc represents the time required to complete the task represented by the arc. For any node in the network, the arcs leading to the node represent the tasks that need to be completed before any task represented by an arc with a tail at that node can be started. A dummy start node s is present in such networks and all tasks which do not need any other task to be completed before they can start have a tail at that node. A dummy stop node t is also present, and all tasks such that the starting of no other task depend on their completion are led to that node.

A longest path from s to t in such networks is called a *critical path* in such *project networks*. Its length denotes the time the project would require to complete, and the tasks represented by the arcs in the critical path are those in which any delay is bound to cause a delay in the completion of the project.

Formally stated, in a *longest path problem*, one is given a weighted network $N = (V, E, w)$ and two nodes s and t, s, $t \in V$. w_e denotes the weight of the edge $e \in E$. A path $P = \{e_1, e_2, \ldots, e_k\}$ is said to have a length of $\sum_{e \in P} w_e$. The required output in the longest path problem is a path P^* between s and t whose length is the maximum among the lengths of all paths between s and t in the network.[2]

1.5.2 The bottleneck shortest path problem

Often when computing a shortest path from a source point to a destination point in logistical problems, one wants to ensure that none of the segments in the path is too long. This is particularly important, for example, while planning routes in sparsely populated regions, where it is very difficult to send repair crews and equipments over long distances. In such situations, one aims to obtain paths for which the length of the longest segment in the path is as small as possible. This problem is commonly referred to as the bottleneck shortest path problem.

Formally stated, in a *bottleneck shortest path problem*, one is given a weighted network $N = (V, E, w)$ and two nodes s, $t \in V$. w_e denotes the weight of the edge $e \in E$. A path $P = \{e_1, e_2, \ldots, e_k\}$ has a bottleneck objective value of $\max_{e \in P} w_e$. The required output in the shortest bottleneck path problem is a path P^* between s and t whose bottleneck objective value is the minimum among all paths between s and t in the network.[3]

1.5.3 The hop-constrained shortest path problem

The hop constrained shortest path problem occurs frequently while optimizing the transmission of signals over a telecommunication network. In a telecommunication network, the nodes represent repeaters. When a signal passes through a repeater, the signal quality deteriorates significantly. Therefore, for a meaningful transmission of

---

[2] For a more detailed treatment of this problem, see H. Kerzner, Project Management: A Systems Approach to Planning, Scheduling, and Controlling, 8th Ed., (2003), Wiley.

[3] The bottleneck shortest path problem has been discussed in O. Berman, D. Einav, G. Handler, The constrained bottleneck problem in networks, Operations Research, 38, (1990), pp. 178–181.

signals, it is often prescribed that the signal should not pass through more than a pre-specified number of repeaters. The optimization problem in such a case becomes one of finding a shortest (or cheapest) way of transmitting a signal from a source node to a destination node, while ensuring that the number of nodes that the signal crosses en-route is no more than a pre-specified number.

Formally stated, in a *hop-constrained shortest path problem*, one is given a weighted network $N = (V, E, w)$, two nodes $s, t \in V$, and a number $k \geq 1$. w_e denotes the weight of the edge $e \in E$. A path $P = \{e_1, e_2, \ldots, e_r\}$ is said to have a length of $\sum_{e \in P} w_e$. The required output in the hop-constrained shortest path problem is a path P^* between s and t having at most $k + 1$ edges whose length is the minimum among the lengths of all paths between s and t having at most $k + 1$ edges.

1.5.4 The Hamiltonian path problem

Consider a situation in which a machine is supposed to operate on several jobs. Each job requires a special setting on the machine, and changing from one job to another requires the machine to be reconfigured. The time required to reconfigure the machine depends on the jobs immediately before and after the reconfiguration. The objective in this problem is to complete all jobs with the minimum total reconfiguration time. If this problem is represented on a network, the nodes of the network correspond to jobs, and the arc from node i to node j has a weight representing the time required to reconfigure the machine from processing job i to processing job j. The problem of finding a shortest path in such a network that passes through each of the nodes is called the Hamiltonian path problem.

Formally stated, in the *Hamiltonian path problem*, one is given a weighted network $N = (V, E, w)$, and two nodes $s, t \in V$. A path $P = \{e_1, e_2, \ldots, e_r\}$ is said to have a length of $\sum_{e \in P} w_e$. The required output in a Hamiltonian path problem is a path P^* between s and t passing through each of the nodes in the network exactly once, whose length is the minimum among the lengths of all such paths in the network.[4]

1.5.5 The stochastic shortest path problem

In all the problem statements in this chapter, we have assumed that the weights of the edges and arcs in the networks are known in advance. In practical situations, this assumption may not be valid. Consider for example, a road network in which the weights on the arcs represent the time taken to traverse the corresponding road segment, and in which a shortest path problem is one of finding the quickest way to go from one particular junction to another. The weights in such a network are stochastic, since the time required to traverse a road segment depends on traffic conditions. Problems of finding appropriately defined best paths in such networks are called *stochastic shortest path problems*.

---

[4] For more details on this problem, see E.L. Lawler, J.K. Lenstra, A.H.G. Rinnooy Kan, D.B. Shmoys, (Eds.), The Traveling Salesman Problem: A Guided Tour of Combinatorial Optimization, 1990, Wiley.

Obviously, since the weights on the arcs are stochastic, the conventional definition of a shortest route is no longer valid; different routes may be shortest with different probabilities. So several objectives are considered in such problems. One may want to set a threshold value of the length of a desirable path, and obtain a path that has length not exceeding this value with the highest probability. Alternatively, one may set a probability value and compute a path that will be shortest with a probability higher than this value.[5]

1.6 Exercises on Shortest Path Problems

The Services Department of GTC manages the relationships between the company and its customers. It handles all interactions with private customers ranging from new subscriptions and connections through maintenance and repair work. Recently, the company has launched the project Gold Service, which includes a 24 hours service. This means that if a customer reports a problem, then a technician makes a site call within 24 hours. Technicians move from one customer to another, and need not go back to the facility where the technician is stationed, except at the end of the working day. The calculation of shortest or fastest routes between two customer sites is not easy in practice, mainly because the circumstances in the road network change rapidly and regularly. Therefore, the company is testing a system which advises a technician about the route he has to take to the next customer, just after he has finished serving a customer.

The road map of the city is schematically depicted in Figure 1.10. There are 50 customer sites denoted by the dots, and labeled 1, ..., 50. A line denotes a direct road connection between two sites. Attached to each road is a number, denoting the distance between the connected sites in kilometers. For instance, the distance between site 22 and site 27 is 3.9 kilometers.

Problem 1.1. Solution by Inspection and Computer
It may be required to obtain a quick idea of distances between arbitrary pairs of sites, for instance if the computer system fails and the calculations need to be done by inspection. Suggest by inspection, as short a route as possible between the following pairs of sites. Do not use more than 20 seconds to determine each separate path.

(1) From site 8 to site 31.
(2) From site 5 to site 38.
(3) From site 1 to site 48.
(4) From site 4 to site 41.
(5) From site 24 to site 30.
(6) From site 9 to site 47.

---

[5] For further reading on this problem, see D.P. Bertsekas, J.N. Tsitsiklis, An analysis of stochastic shortest path problems, Mathematics of Operations Research 16, (1991), pp. 580–595.

Now use a computer program to calculate shortest routes (in kilometers) between the pairs of sites given above. It is normal that the computer calculations yield better results.

In order to obtain an idea about the time required to go from one location to another, we can also make use of a road map in which the travel times are given for each connected pair of locations. We have depicted the schematic road map in Figure 1.11 with the travel times in minutes. Travel times are of course less precisely determined than distances, and may also change suddenly. These uncertainty aspects have to be taken into account when performing realistic calculations based on them.

Problem 1.2. What Happens If

Use the time network depicted in Figure 1.11 to respond to the following. Unless specified each part is independent of the others.

(a) Calculate quickest routes (in minutes) between the pairs of sites from Problem 1.1.
(b) Determine for each of these quickest routes the range within which the travel time between the sites 8 and 10 may change without changing the calculated quickest route.
(c) What could be a plausible reason for the range in (b) in the case of Problem 1.1 pair (1) being much larger than the one for Problem 1.1 pair (2)?
(d) Show, using tolerance intervals, that the shortest route in case of Problem 1.1 pair (6) is not unique. Calculate an alternative optimal solution. Answer the same questions for Problem 1.1 pair (3).
(e) Due to an accident, no traffic is possible between the sites 8 and 10. Show that the length of any shortest route that contains road segment 8 – 10 is changed as if the value of road segment 8 – 10 is pegged to its upper tolerance value with respect to that quickest route.
(f) Determine a second quickest route between 8 and 31 for the network in Figure 1.11. Show that the "length" (in minutes) of this route can also be determined by using tolerance intervals.
(g) Quickest routes sometimes need to be determined in the presence of restrictions that certain roads of the network have to be included (or excluded) in the route. Such problems may occur when cable work activities are executed in certain streets and the work in process has to be checked. Determine quickest routes in the following situations.

1. From site 27 to site 16, including road segment 21 – 22.
2. From site 35 to site 26, including road segment 31 – 34.
3. From site 39 to site 3, including road segment 16 – 17.
4. From site 44 to site 14, including road segment 27 – 28.
5. From site 15 to site 14, including road segments 16 – 10 and 12 – 22.
6. From site 26 to site 48, including road segments 35 – 40 and 40 – 41. Determine a second best and a third quickest route as well.

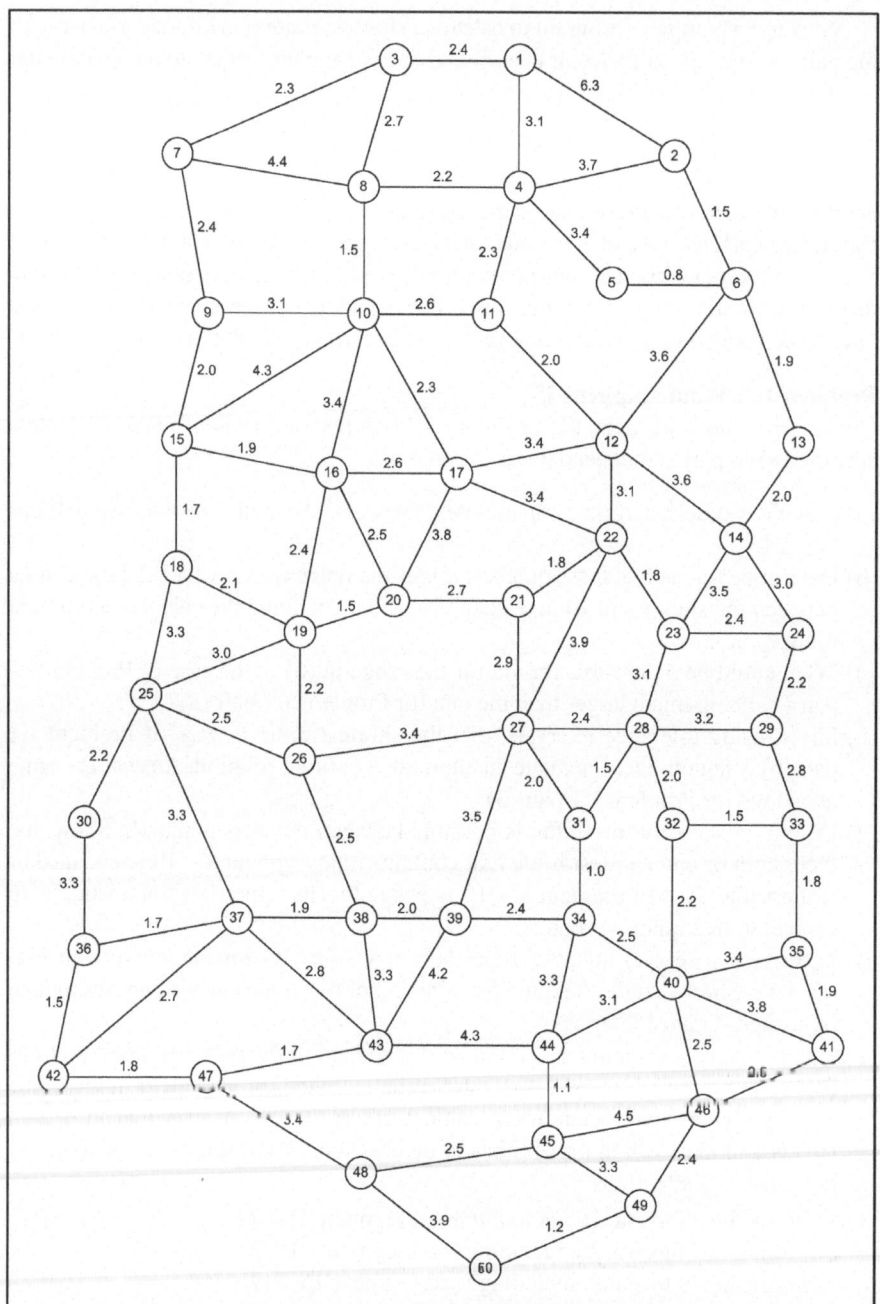

Fig. 1.10. Schematic road map with distances in kilometers

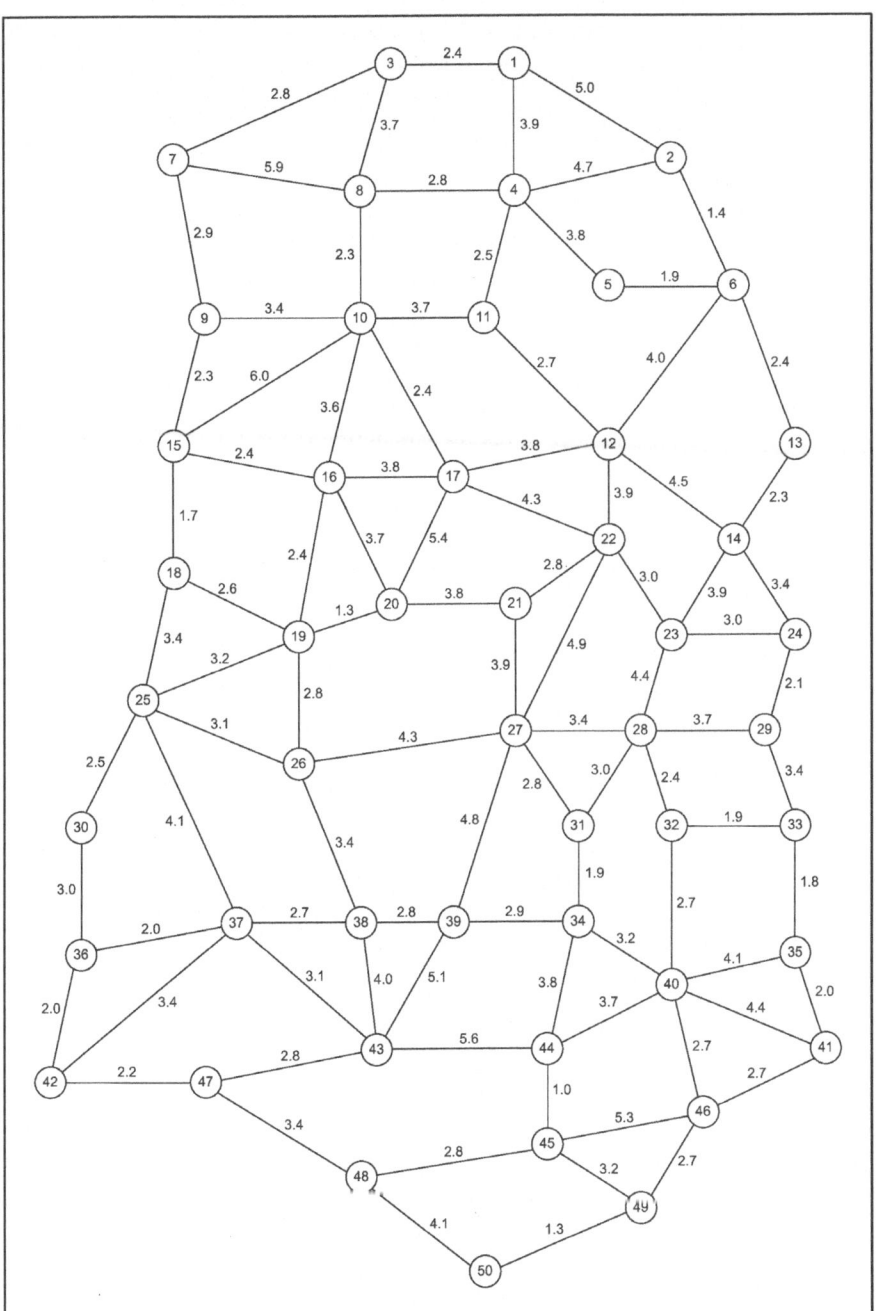

Fig. 1.11. Schematic road map with travel times in minutes

7. A driver goes from site 37 to site 43 and has to pass the road segments 35 – 40 and 40 – 41. Is this possible to do within 35 minutes?

The management of GTC has decided to replace its old cable network. The Research & Development (R&D) department has been asked to design a new cable that has ten times more capacity than the old cable. The cable has to be developed within the next five years. In this period all the three activities, viz. actual research, testing, and evaluation, have to be carried out. These three activities will be executed simultaneously. The most expensive part of the project concerns wages and training of the research employees. The company has already employed a number of researchers, but during the five years extra researchers will be needed. These extra researchers need to undergo special training due to the specific character of the work. R&D requires a lower number of researchers during the testing period than during the evaluation and research periods. It has decided to work with half-yearly planning periods because, in general, the nature of the work changes noticeably every half year. The company has to design a schedule of hiring and firing people. The next question is on determining an optimal hiring-firing schedule.

Problem 1.3. Manpower Planning
The numbers of extra research employees needed for the research and development of the new cable for the next ten consecutive half-year periods are listed in Table 1.1. The half-year periods are labeled $1, \ldots, 10$. The training of a new researcher requires an investment of €8,000. Firing a person is expensive, it costs €24,000 per fired person. Moreover, if there are fewer people under contract than required, it costs the company an average of €32,000 per person per half-year.

Table 1.1. Extra researcher requirement for the project

Period	1 2 3 4 5 6 7 8 9 10
Extra researchers	9 7 6 12 11 8 13 11 7 10

(a) Determine a hiring-firing schedule to minimize cost, such that the number of employed researchers is as high as possible during each period. Use a shortest path problem formulation for solving this problem. What is the total number of redundant researchers in your solution?
(b) In the periods 3 and 6 the expected work pressure is lower than normal. Determine an optimal solution for which the number of researchers under contract is minimal during these periods.

Problem 1.4. Subsidizing Connections

In order to test the new cable, the company has decided to connect the university, located at site 3 of the network of Figure 1.10, to the R&D building of GTC, located at site 46, using its cable network.

The City Council desires that two clinics associated with the city hospital, which are located at sites 8 and 20, be connected by a high capacity cable. So it wants to find out how much subsidy to give to GTC such that it is profitable for GTC to connect the two sites 8 and 20 with the new cable.

The City Council has decided to subsidize the laying of cable on the paths 8 – 10 – 16 – 20 and 29 – 28 – 31 – 34. The cables on either of the two paths cannot be interrupted. In order to minimize the disruption to the traffic circulation, it has also decided that work on laying cable on the paths 8 – 10 – 16 – 20, 16 – 19 – 26 – 38 – 37 – 42, 29 – 28 – 31 – 34 – 39 – 27 – 22 – 12 – 6, and 48 – 45 – 44 – 40 – 32 – 33 has to be executed in the directions $8 \to 10 \to 16 \to 20$, $16 \to 19 \to 26 \to 38 \to 37 \to 42$, $29 \to 28 \to 31 \to 34 \to 39 \to 27 \to 22 \to 12 \to 6$, and $48 \to 45 \to 44 \to 40 \to 32 \to 33$, respectively. Actually, the subsidy also yields some extra profit for GTC. The construction costs are a multiple of the distances from Figure 1.10, in €1,000's per km. The subsidies on road segment $8 \to 10$ is €3,500, on road segment $29 \to 28$ is €6,800, on road segment $10 \to 16$ is €6,600, on road segment $28 \to 31$ is €2,900, on road segment $16 \to 20$ is €4,900, and on road segment $31 \to 34$ is €2,200.

(a) Is it true that the subsidy is high enough, i.e., is it profitable for GTC to connect the sites 3 and 46 via the subsidized paths?

Knowing that GTC is also considering a new cable between the sites 2 and 50, and having confidence in a positive effect of the subsidies stated above, the City Council has approved an extra subsidy on paths 45 – 44 – 40 – 32 and 27 – 22 – 12 – 6. The subsidy on road segment $40 \to 32$ is €3,200, on road segment $12 \to 6$ is €7,700, on road segment $44 \to 40$ is €4,700, on road segment $22 \to 12$ is €7,100, on road segment $45 \to 44$ is €2,800, and on road segment $27 \to 22$ is €6,900.

(b) Explain why it is not possible to use a shortest path algorithm to decide on the profitability for GTC to lay the cable on the two paths.
(c) After a re-calculation the City Council has decided to decrease the subsidy on 29 – 28 from €6,800 to €5,500. Answer the same question as in part (b).
(d) Determine a solution (need not be optimal) for the following situation. The sites 3 and 46, and the sites 2 and 50 need to be connected by a cable, one between 3 and 46, and one between 2 and 50. If the two cables connect two sites with subsidy, then the subsidy is given only once.

2

Minimum Spanning Trees

2.1 Introduction

GTC has been assigned the job of interconnecting six departments, labeled A, B, C, D, E, and F, of a university, at minimum cost. Practical considerations make it impossible to connect several pairs of departments directly to one another. In fact the only direct connections possible are the ones between departments A and B, A and D, B and C, B and D, B and E, C and D, C and F, D and E, and E and F. These connections are shown in the network of Figure 2.1, in which the nodes correspond to departments and edges correspond to possible direct connections. The numbers next to the edges denote the cost of making those connections in €100 units. GTC is required to connect the six departments as cheaply as possible.

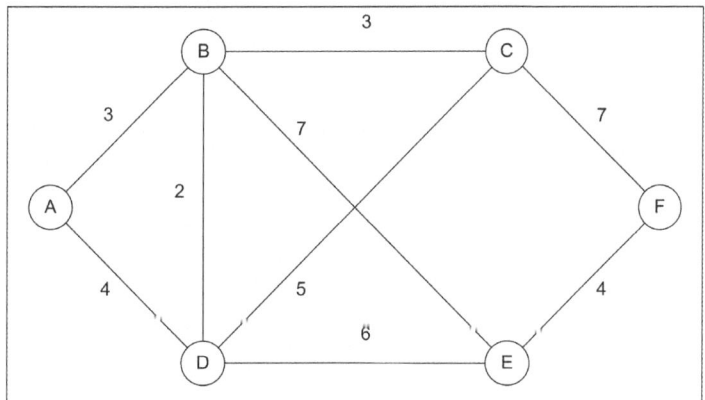

Fig. 2.1. Possible direct connections among the departments

It is easy to predict some of the properties of the network solution that GTC would come up with. First, the network solution obviously needs to connect all six

G. Sierksma and D. Ghosh, *Networks in Action: Text and Computer Exercises in Network Optimization*, International Series in Operations Research & Management Science 140, DOI 10.1007/978-1-4419-5513-5_5, © Springer Science + Business Media, LLC 2010

departments, i.e., it needs to span all the nodes in the network of Figure 2.1. Second, there should not be more than one path between any pair of departments in the network solution, i.e., it should be acyclic. This means for example, that all of the four edges A – B, B – C, C – D, and A – D should not be simultaneously present in the solution. These four edges together link the four departments, A, B, C, and D, at a cost of €1500. However, if any one of the four, say C – D, is removed from the solution, the four departments still remain connected, but the connection cost reduces to €900. A network with no cycles is called a *tree*, and so the network solution that GTC suggests has to be a *spanning tree*, i.e., an acyclic subset of edges linking all the nodes in the network. Since the objective is to obtain a minimum cost solution, GTC needs to find a minimum spanning tree, which is a spanning tree in the network, the sum of whose connection costs is the minimum possible. One such network solution is shown in Figure 2.2. The thicker lines in the figure represent the connections that

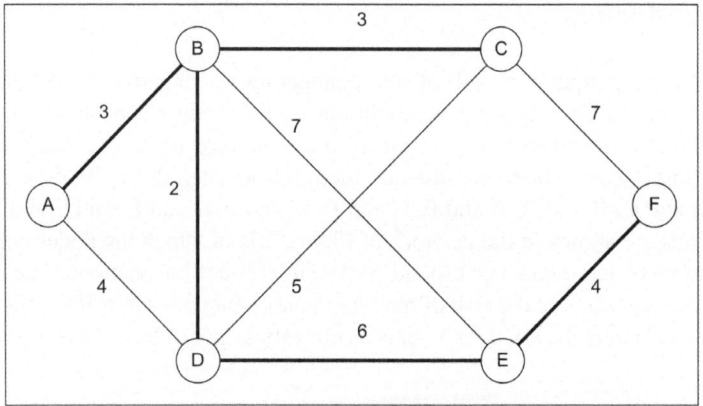

Fig. 2.2. A network solution (a minimum spanning tree)

are present in the network solution. The total connection cost for this network solution is €1800. This chapter deals with the construction of such minimum spanning trees.

Formally stated, in a *minimum spanning tree*, one is given a connected weighted network $N = (V, E, w)$ where each of the entries w_e of w denote the weight associated with the edge $e \in E$. A spanning tree T in the network is a subset of edges of N such that no subset of edges in T form a cycle, and for each node in the network, there is an edge in T that is incident on it. The weight of the tree T is the sum $\sum_{e \in T} w_e$. One is required to output a spanning tree in N with the minimum weight.

Notice that a minimum spanning tree is *not* the union of edges of the shortest paths between all pairs of nodes in a network. For example, in the solution shown in Figure 2.2, the cost of the connection between the departments C and F is €1500, although a direct connection between them is possible at a cost of €700. The reader can verify that the connections in a spanning tree forced to contain the edge C – F will cost at least €1900.

Also notice that a minimum spanning tree in a network may not be unique. In the example above, the minimum spanning tree is unique. However, if a network has a pair of equal length paths between a pair of nodes, then that network may have multiple minimum spanning trees; see e.g., Figure 2.3.

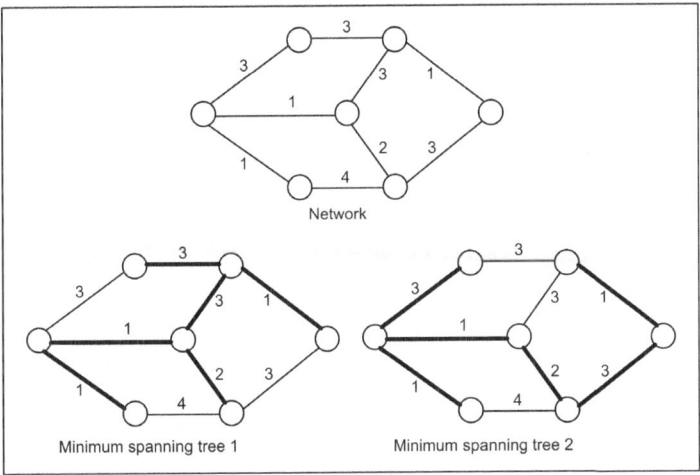

Fig. 2.3. Network with multiple minimum spanning trees

2.2 Applications

Minimum spanning trees have a variety of applications. Some of them are described below.

2.2.1 Designing networks for disasters

Consider a situation in which the transportation network in an area has been completely damaged due to some natural disaster, like a flood. In order to supply essential commodities to the population centers in the area, a part of the road network needs to be developed as soon as possible. A design problem in such cases is to find out how to connect the population centers to each other and to a depot, so that essential commodities can be routed from the depot to the population centers. Since construction takes time, the problem reduces to one of finding a minimal network linking all the population centers to the depot. This problem can be modeled as a minimum spanning tree where the nodes in the network are the population centers and the depot, the edges are possible connections between the population centers, and the weights on the edges represent the distance between the population centers that the edge connects.

2.2.2 Group technology

Group technology is a manufacturing philosophy that takes into account the similarities in parts and assemblies of a group of products in order to classify them into families. Products in a single family share commonalities in design, manufacturing, and other design processes. One important problem in group technology is to decide on the partitioning of products into families. This problem can be modeled using minimum spanning trees. A network of the products is constructed as follows. Each product corresponds to a node in the network. The dissimilarities between a pair of products is quantified and is represented in the network as the cost of the edge joining the nodes corresponding to the products. The more dissimilar the products are, the larger is the cost of the corresponding edge. A minimum spanning tree for the network thus obtained is then constructed. The part families are then generated by removing an appropriate number of the high cost edges in the minimum spanning tree.

2.2.3 Storing large but similar data

In many studies in genetics, one needs to store large amounts of data in the most efficient manner possible. The different elements of data are voluminous, but similar to each other, and vary from one another only at a few places. A common way of representing such data is to store one data element in its entirety and store others by just keeping track of how they differ from another stored data element. Of course, an important problem here is to find out the sequence of storage of the different data elements, so that the total storage is minimized. This problem can be solved using minimum spanning trees. Consider a network where each node corresponds to one element. Each pair of nodes in the network is linked using edges. The cost of each such edge is the amount by which the data corresponding to the two nodes differ. A minimum spanning tree is then constructed on this network and one node is chosen as the reference node. The element corresponding to this node is the reference element, and the data corresponding to this element is written out in full. Each neighbor of this element is next considered and the data for the element are written down by first pointing to the reference data element, and then noting down the difference from that element. Continuing in this way, data for each element are stored. In order to retrieve data about an arbitrary element, one needs to start at the reference node, and compute a path to the element. Then one needs to note the data corresponding to the reference element, and then make changes in the data as prescribed by nodes along the path until one reaches the element of interest.

2.3 Linear Programming Formulations

There are many different linear programming formulations for the minimum spanning tree problem. In this section two intuitively appealing formulations are presented. Both formulations naturally occur as integer linear programs, but relaxing the requirement that some of the decision variables need to be integers in these formulations does not affect their optimal solutions.

In the first formulation, a decision variable x_{ij} is defined for each edge $i - j$ in E. For any edge $i - j$ in E, $x_{ij} = 1$ in an optimal solution implies that $i - j$ is included in the minimum spanning tree output, and $x_{ij} = 0$ implies that it is not. The objective is to generate a spanning tree with minimum cost, i.e., to minimize the sum of the weights of the edges included in the solution. Hence the objective is mathematically expressed as:

$$\text{Minimize} \sum_{i-j \in E} w_{ij} x_{ij}. \tag{2.1}$$

The formulation needs to constrain the solution set to include only spanning trees in the network. A spanning tree in a network with n ($n \geq 1$) nodes has exactly $n - 1$ edges. So the formulation includes a constraint that the number of edges in the minimum spanning tree output by the formulation has $|V| - 1$ edges. This is achieved by specifying the constraint:

$$\sum_{i-j \in E} x_{ij} = |V| - 1. \tag{2.2}$$

A second condition stipulates that no cycles should be present in the set of edges that define a minimum spanning tree. The formulation implements this by ensuring that for each subset S of V, the number of edges in the solution output that are members of S is strictly less than $|S|$, where $|S|$ denotes the number of elements of S. This is implemented using the following set of constraints:

$$\sum_{i-j \in E;\ i,j \in S} x_{ij} \leq |S| - 1 \qquad \text{for each } S \subseteq V. \tag{2.3}$$

When $|S| = 1$, constraint set (2.3) is meaningless, since in the networks considered here, there are no loops (edges that connect a node to itself). When $|S| = 2$, constraint set (2.3) implies that no x_{ij} value can exceed 1. Further, for any set $S \subset V$, with $|S| \geq 3$, constraint set (2.3) is meaningless unless there are at least $|S|$ edges with both ends in S. Again when $S = V$, constraint set (2.3) reduces to constraint (2.2). Thus, in any reasonably sparse network, only a small subset of the $2^{|V|}$ constraints of the constraint set (2.3) is required.

Due to a special structure of the constraint coefficients called *total unimodularity*, it can be shown that relaxing the condition that "each x_{ij} can attain only values of 0 or 1" to the condition that "each x_{ij} must lie between 0 and 1" does not affect the values of x_{ij}'s in any optimal solution to the formulation. Furthermore, in the relaxed problem, the condition that $x_{ij} \leq 1$ need not be added, since constraint (2.3) for subsets of V with two nodes ensure that x_{ij} cannot exceed 1.

The first formulation to obtain a minimum spanning tree in a network $N = (V, E, w)$ is shown in its entirety in Figure 2.4. As an illustration of the formulation, Figure 2.5 shows the linear programming formulation for the network in Figure 2.1. Notice that the number of constraints in this formulation is small compared to the $2^6 + 1$ constraints prescribed in the general formulation in Figure 2.4.

Minimize

$$z = \sum_{i-j \in E} w_{ij} x_{ij}$$

Subject to

$$\sum_{i-j \in E} x_{ij} = |V| - 1$$

$$\sum_{\substack{i-j \in E \\ i,j \in S}} x_{ij} \leq |S| - 1 \quad \text{for each } S \subset V,\, S \neq \emptyset$$

$$x_{ij} \geq 0 \qquad \text{for each } i - j \in E$$

Fig. 2.4. First linear programming formulation of the minimum spanning tree problem

The second formulation is closely related to the first. Given a network $N = (V, E, w)$, here too, for each edge $i - j$ in E, a variable x_{ij} is defined. If $x_{ij} = 1$ in an optimal solution, then edge $i - j$ is included in the optimal solution output, and if $x_{ij} = 0$, then it is not. As in the first formulation, the objective is to minimize the weight of the tree output; hence the objective of the formulation is:

$$\text{Minimize} \sum_{i-j \in E} w_{ij} x_{ij}. \tag{2.4}$$

The number of edges included in any optimal solution is $|V| - 1$, so the first constraint in the formulation is again:

$$\sum_{i-j \in E} x_{ij} = |V| - 1. \tag{2.5}$$

This formulation differs from the first formulation in the way it avoids cycles. Assume a set T containing $|V| - 1$ edges from E in the network N. It is intuitively obvious and also easy to show that if and only if T contains a cycle, there would be at least two nodes in V which will not be connected to each other by an edge in T. The current formulation makes use of this result to avoid cycles. It arbitrarily chooses a node $s \in V$, constructs all proper subsets S of V containing s, and for each S, constrains at least one edge in any feasible solution to connect a node in S to a node outside S. It achieves this using the constraint set:

$$\sum_{\substack{i-j \in E \\ i \in S, j \notin S \text{ or} \\ i \notin S, j \in S}} x_{ij} \geq 1 \text{ for each } S \subset V \text{ such that } s \in S. \tag{2.6}$$

The second formulation to obtain a minimum spanning tree in a network $N = (V, E, w)$ is shown in its entirety in Figure 2.6.

Minimize

$$z = 3x_{AB} + 4x_{AD} + 3x_{BC} + 2x_{BD} + 5x_{BE} + 5x_{CD} + 7x_{CF} + 6x_{DE} + 4x_{EF}$$

Subject to

$$x_{AB} + x_{AD} + x_{BC} + x_{BD} + x_{BE} + x_{CD} + x_{CF} + x_{DE} + x_{EF} = 5 \quad \text{(Constraint (2.2))}$$

$$x_{AB} \leq 1 \quad \text{(Constraint (2.3), S} = \{A, B\})$$
$$x_{AD} \leq 1 \quad \text{(Constraint (2.3), S} = \{A, D\})$$

There are seven more similar constraints in which $|S| = 2$.

$$x_{AB} + x_{AD} + x_{BD} \leq 2 \quad \text{(Constraint (2.3), S} = \{A, B, D\})$$
$$x_{BC} + x_{BD} + x_{CD} \leq 2 \quad \text{(Constraint (2.3), S} = \{B, C, D\})$$
$$x_{BD} + x_{BE} + x_{DE} \leq 2 \quad \text{(Constraint (2.3), S} = \{B, D, E\})$$

$$x_{AB} + x_{AD} + x_{BC} + x_{BD} \leq 3 \quad \text{(Constraint (2.3), S} = \{A, B, C, D\})$$
$$x_{AB} + x_{AD} + x_{BD} + x_{DE} \leq 3 \quad \text{(Constraint (2.3), S} = \{A, B, D, E\})$$

There are five more similar constraints in which $|S| = 4$.

$$x_{AB} + x_{AD} + x_{BC} + x_{BD} + x_{BE} + x_{CD} + x_{DE} \leq 4 \quad \text{(Constraint (2.3), S} = \{A, B, C, D, E\})$$
$$x_{AB} + x_{AD} + x_{BC} + x_{BD} + x_{CD} + x_{CF} \leq 4 \quad \text{(Constraint (2.3), S} = \{A, B, C, D, F\})$$

There are four more similar constraints in which $|S| = 5$.

$$x_{AB}, x_{AD}, x_{BC}, x_{BD}, x_{BE}, x_{CD}, x_{CF}, x_{DE}, x_{EF} \geq 0 \quad \text{(Non-negativity constraints)}$$

Fig. 2.5. First formulation of the minimum spanning tree problem for the network in Figure 2.1

As an illustration, the second formulation for obtaining a minimum spanning tree in the network in Figure 2.1 is shown in Figure 2.7. In this case, the node s is the node A in the network.

2.4 Algorithms for Minimum Spanning Trees

There are several specialized algorithms for obtaining minimum spanning trees in networks, but most of them depend on the following result.

Assume that we have a connected weighted network, and have a set of edges in the network that we know are part of some minimum spanning tree on that network. Let these edges and the nodes that they are incident on define

Minimize

$$z = \sum_{i-j \in E} w_{ij} x_{ij}$$

Subject to

$$\sum_{i-j \in E} x_{ij} = |V| - 1$$

$$\sum_{\substack{i-j \in E, \\ i \in S, j \notin S \text{ or} \\ i \notin S, j \in S}} x_{ij} \geq 1 \qquad \text{for each } S \subset V \text{ such that } s \in S$$

$$x_{ij} \in \{0,1\} \quad \text{for each } i - j \in E$$

Fig. 2.6. Second linear programming formulation of the minimum spanning tree problem

components of the network. Then for each of the components, the least cost edge that connects the component to any of the other components will be part of at least one minimum spanning tree in the network.

For example, consider Figure 2.8. Assume that we known that the subset of edges A – D, B – E, and E – G are present in a minimum spanning tree in the network. These edges define four components in the network, shown in the bottom right hand diagram in Figure 2.8. Consider the component formed by the edges B – E and E – G, and nodes B, E, and G. There are three edges, namely A – B, D – E, and F – G, that connect this component to other components in the network. Of these, edge A – B has the lowest cost. So according to the result, there is at least one minimum spanning tree on this network that contains the edge A – B, in addition to the edges A – D, B – E, and E – G.

We will next present two popular algorithms for finding minimum spanning trees that are based on the result described above.

2.4.1 Prim's algorithm

The algorithm that we describe here is due to R.C. Prim (1957), and is known as Prim's algorithm.

Consider a weighted connected network $N = (V, E, w)$ ($|V| = n \geq 2$). Prim's algorithm starts by constructing another network $N' = (V, T, w)$ where $T - \emptyset$. Clearly then, N' has n components. Prim's algorithm chooses one of the components of N' at random and uses the result, described earlier, on this component. To do this, it considers all edges in E that connect the component to other components in N' and chooses the least cost edge among these. This edge is then added to T and the first iteration is over.

Minimize

$$z = 3x_{AB} + 4x_{AD} + 3x_{BC} + 2x_{BD} + 5x_{BE} + 5x_{CD} + 7x_{CF} + 6x_{DE} + 4x_{EF}$$

Subject to

$$x_{AB} + x_{AD} + x_{BC} + x_{BD} + x_{BE} + x_{CD} + x_{CF} + x_{DE} + x_{EF} = 5 \quad \text{(Constraint (2.5))}$$

$$x_{AB} + x_{AD} \geq 1 \quad \text{(Constraint (2.6), S = \{A\})}$$

$$x_{AD} + x_{BC} + x_{BD} + x_{BE} \geq 1 \quad \text{(Constraint (2.6), S = \{A, B\})}$$
$$x_{AB} + x_{BD} + x_{CD} + x_{DE} \geq 1 \quad \text{(Constraint (2.6), S = \{A, D\})}$$

There are three more similar constraints in which $A \in S$ and $|S| = 2$.

$$x_{AD} + x_{BD} + x_{BE} + x_{CD} + x_{CF} \geq 1 \quad \text{(Constraint (2.6), S = \{A, B, C\})}$$
$$x_{BC} + x_{BE} + x_{CD} + x_{DE} \geq 1 \quad \text{(Constraint (2.6), S = \{A, B, D\})}$$

There are eight more similar constraints in which $A \in S$ and $|S| = 3$.

$$x_{BE} + x_{CF} + x_{DE} \geq 1 \quad \text{(Constraint (2.6), S = \{A, B, C, D\})}$$
$$x_{BC} + x_{CD} + x_{EF} \geq 1 \quad \text{(Constraint (2.6), S = \{A, B, D, E\})}$$

There are eight more similar constraints in which $A \in S$ and $|S| = 4$.

$$x_{CF} + x_{EF} \geq 1 \quad \text{(Constraint (2.6), S = \{A, B, C, D, E\})}$$
$$x_{DE} + x_{EF} \geq 1 \quad \text{(Constraint (2.6), S = \{A, B, C, D, F\})}$$

There are three more similar constraints in which $A \in S$ and $|S| = 5$.

$$x_{AB}, x_{AD}, x_{BC}, x_{BD}, x_{BE}, x_{CD}, x_{CF}, x_{DE}, x_{EF} \in \{0, 1\} \quad \text{(Integrality constraints)}$$

Fig. 2.7. Second formulation of the minimum spanning tree problem for the network in Figure 2.1

At each of the subsequent iterations, Prim's algorithm chooses the component of N' that has the largest cardinality and applies the result on it to pick another edge to include in T. Since this procedure increases the cardinality of the chosen component by 1 but does not alter the cardinality of any of the other components, Prim's algorithm is said to "grow" a tree in the network, which becomes a minimum spanning tree at the end of the algorithm's execution.

An implementation of Prim's algorithm works by creating and maintaining three sets called ADDED, TO-ADD, and TREE. ADDED stores the nodes in the component of N' that Prim's algorithm decides to augment at each iteration. TO-ADD contains the nodes in all other components of N'. TREE contains the edges that have

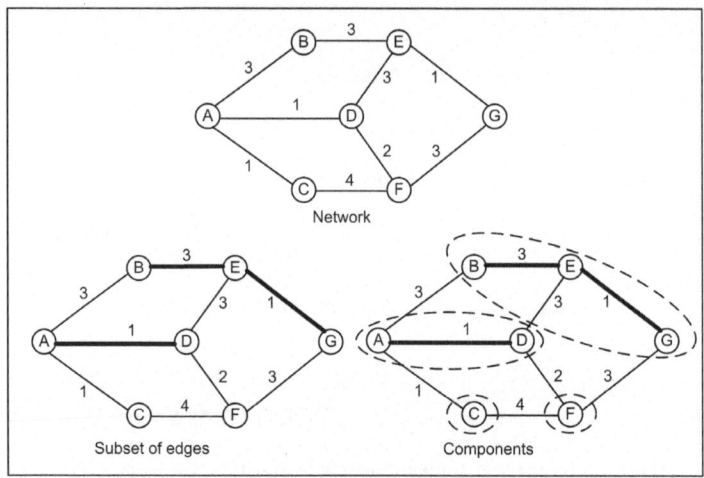

Fig. 2.8. Components in a network

been added to T. We initialize ADDED and TREE to empty sets and add all nodes in the network to TO-ADD. In the course of the algorithm, we will deplete the TO-ADD set and populate the ADDED and TREE sets. We choose any node from TO-ADD and move it to ADDED. In each step of the algorithm, we choose a node u from TO-ADD and v from ADDED, such that the cost of the connection between u and v is the minimum among all the connections between nodes in TO-ADD and ADDED. (If there are ties, then these are broken arbitrarily, thus leading to alternate minimum spanning trees.) Then we remove u from TO-ADD and add it to ADDED, and add the edge $u - v$ to TREE. For a connected network containing n nodes, the algorithm stops after $n - 1$ iterations, (note that in a network with n nodes, any minimum spanning tree has exactly $n - 1$ edges), and the edges in TREE define a minimum spanning tree for the network.

Let us illustrate Prim's algorithm on the network in Figure 2.1. Initially ADDED and TREE are empty sets, while TO-ADD = {A, B, C, D, E, F}. We arbitrarily put A in the ADDED set in the beginning of the first iteration. Then we will choose B as the node u, A as the node v in the first iteration since A – B has the least cost among all edges connected to nodes in ADDED. At the end of this iteration, ADDED = {A, B}, TO-ADD = {C, D, E, F}, and TREE = {A – B}. In the next iteration, the algorithm considers all edges connecting nodes in ADDED to nodes in TO-ADD. Therefore, it considers the edges A – D, B – C, B – D, and B – E. Of these the lowest cost edge is B – D. Therefore, the algorithm adds B – D to TREE, and moves node D from TO-ADD to ADDED. The algorithm proceeds and stops after five iterations. The edges in TREE at the end of each iteration are shown in Figure 2.9.

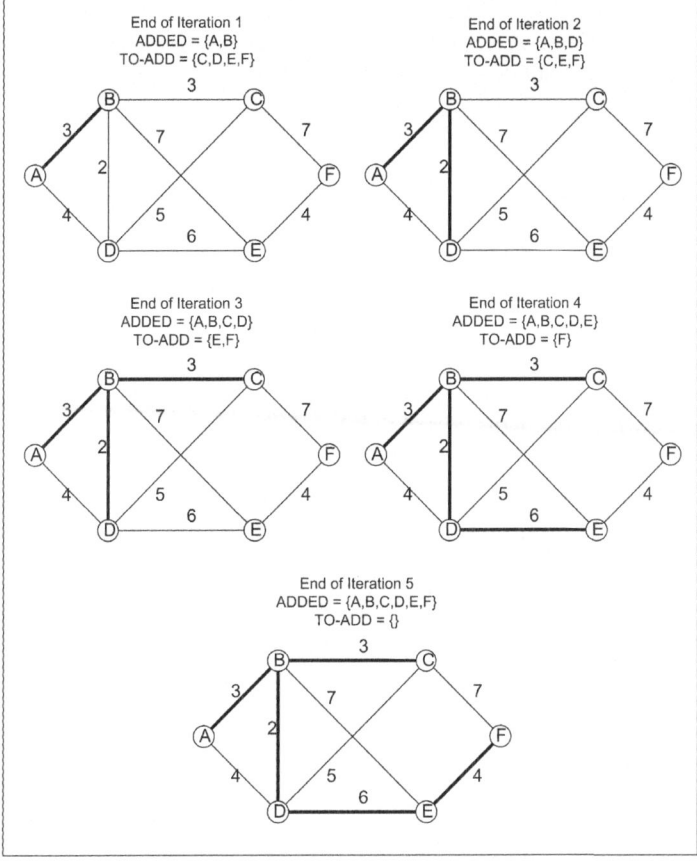

Fig. 2.9. Prim's algorithm in action

2.4.2 Kruskal's algorithm

Kruskal's algorithm, due to J. Kruskal (1956), uses the same ideas as used in Prim's algorithm, but applies it in a slightly different way. Instead of growing a single tree, the algorithm grows a forest, and finally combines them together to form a minimum spanning tree.

Consider a weighted connected network $N = (V, E, W)$ ($|V| = n \geq 2$), on which Kruskal's algorithm is to be run. Kruskal's algorithm starts by constructing another network $N' = (V, T, W)$ where $T = \emptyset$. Here too, since $T = \emptyset$, N' has n components. At each iteration, Kruskal's algorithm chooses the minimum cost edge in E that connects two different components in N' and includes it in T. As a result, the two components that the edge is adjacent to coalesce into one component at the end of the iteration. This algorithm generates a minimum spanning tree because of the following reason. Let e be the edge that is chosen to be included in T at a particular iteration. Let e be incident on components C_1 and C_2 in N'. Since e is a least cost edge that joins any two components at that iteration, it is surely a least cost edge that joins C_1 (or C_2) to any of the other components. So according to the result stated earlier, e must be in a minimum spanning tree.

Implementations of Kruskal's algorithm maintain a list of edges in the network N sorted in non-increasing order of costs. We call this list LIST. They also maintain a list CMP of components of the network N' at each iteration, and a set TREE that would contain the edges in the minimum spanning tree at the termination of the algorithm. Initially, LIST contains all the edges in E arranged in non-increasing order of costs, and TREE is empty. Since TREE is empty, each of the nodes in N form one component in CMP. At each iteration, a lowest cost edge in LIST is removed from it. If the endpoints of the edge are in different components in CMP, then the edge is added to TREE. Else the next least cost edge is chosen from LIST. The iteration is over when one edge has been added to TREE.

Let us now illustrate Kruskal's algorithm on the network in Figure 2.1. Initially, TREE is empty, LIST = {A − B, A − D, B − C, B − D, B − E, C − D, C − F, D − E} and CMP = {{A}, {B}, {C}, {D}, {E}, {F}}. In the first iteration, B − D is taken out of LIST. Since the endpoints of B − D, i.e., nodes B and D are in different components in CMP, B − D is added to TREE, and component {B, D} replaces components {B} and {D} in CMP. In the next iteration, A − B is added to TREE and component {A, B, D} replaces components {A} and {B, D} in CMP. The algorithm stops after five iterations. Figure 2.10 depicts the contents of LIST and CMP at the end of each iteration. The edges marked with thick lines are the ones included in TREE at the end of the iteration.

2.5 Other Tree Problems

2.5.1 The Steiner tree problem

Consider a situation where the transportation links to a set of locations have been destroyed by a natural calamity. In order to make rescue operations most effective, one is required to connect the locations with large populations to a central relief camp. However, in order to construct makeshift transportation networks, it may be cost effective to route them through locations with small populations, simply because such locations are favorably located. This problem of using intermediate locations in a network to connect a pre-specified set of locations with a minimum cost tree is called the Steiner tree problem.

Formally stated, in a *Steiner tree problem*, one is given a network $N = (V, E, w)$ and a subset of nodes V_s. The objective in the problem is to output a tree spanning all nodes of V_s (and some nodes in $V \setminus V_s$ if required) such that the sum of the weights of the edges of the tree is the minimum possible.[1]

---

[1] For more details on this problem, see F.K. Hwang, D.S. Richards, P. Winter, The Steiner Tree Problem, 1992, North Holland.

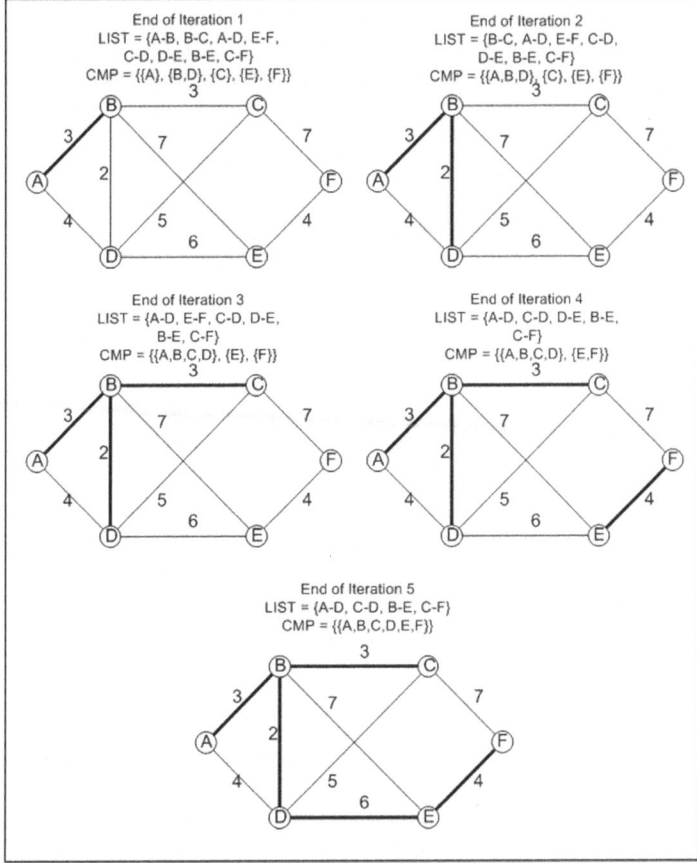

Fig. 2.10. Kruskal's algorithm in action

2.5.2 The capacitated minimum spanning tree problem

Consider a wired computer network in which several "dumb" terminals need to be connected to a central computer. In order to minimize costs, these terminals should be connected in a spanning tree topology. Each tree of nodes branching off from the central computer is connected to the central computer at a single port. Since these terminals exchange data with the central computer, and since a port of a computer has a fixed capacity, while constructing the spanning tree, one should ensure that the number of dumb terminals connected to a single port, i.e., the size of any subtree branching from the central computer should not exceed a fixed value. A spanning tree that has a restriction that none of the subtrees that branch from a pre-specified node has more than a fixed number of nodes is called a capacitated spanning tree. So the problem of designing such a computer network is called the capacitated minimum spanning tree problem.

Formally stated, in a *capacitated minimum spanning tree problem*, one is given a network $N = (V, E, w)$, a node $v \in V$, and an integer k, and one is required to output a minimum cost tree spanning V such that none of the subtrees rooted at v has more than k nodes in it.[2]

2.5.3 The degree constrained minimum spanning tree problem

Consider a wired computer network in which the computers themselves act as routers. If one wants to build such a network at minimum cost, then the topology that one would choose would be a spanning tree. However, computer failures can affect the network severely, for example, if a computer is directly connected to r ($r \geq 1$) other computers, then if it fails, the network is immediately broken up into r sub-networks which cannot communicate with each other. In order to minimize this kind of a problem, one can consider building minimum spanning tree networks in which each node cannot be connected to more than a pre-specified number of other nodes. The design of this type of networks is called a degree constrained minimum spanning tree problem.

Formally stated, in a *degree constrained minimum spanning tree problem*, one is given a network $N = (V, E, w)$ and a positive integer k, and one is required to output a spanning tree in which no node has a degree more than k, and the sum of the weights of the edges of the tree is as small as possible.[3]

2.5.4 The most reliable minimum spanning tree problem

In critical situations, it is more important that all nodes in a network stay connected than finding a minimum cost connection. In such cases, the problem is to output, given a network $N = (V, E, w)$ and the probability of each edge failing, a spanning tree that has the highest probability of remaining intact. This problem is called the *most reliable minimum spanning tree problem*.

2.6 Exercises on Minimum Spanning Tree Problems

In a project similar to the one described in the introduction to this chapter, GTC has been provided with 50 locations to connect in a certain country. These locations need to be connected in such a way that any two of these locations are able to communicate with each other. All the connections need not be direct. Due to high bandwidth requirements, GTC will use high capacity cables for the connections, and wants to know the costs of laying the cable under various circumstances. In Figure 2.11 we have schematically depicted the 50 locations and all possible direct connections. The numbers attached to the connections in Figure 2.11 refer to the distances (in 10 kilometer units) between locations. The cost of the cable is €5,000 per kilometer.

---

[2] For more details on this problem, see K. Chandy and T. Lo, The capacitated minimal spanning tree problem, Networks 3, (1973), pp.173–182.

[3] For more details on this problem, see S.C. Narula and C.A. Ho, Degree constrained minimum spanning tree, Computers & Operations Research 7, (1980), pp.239–249.

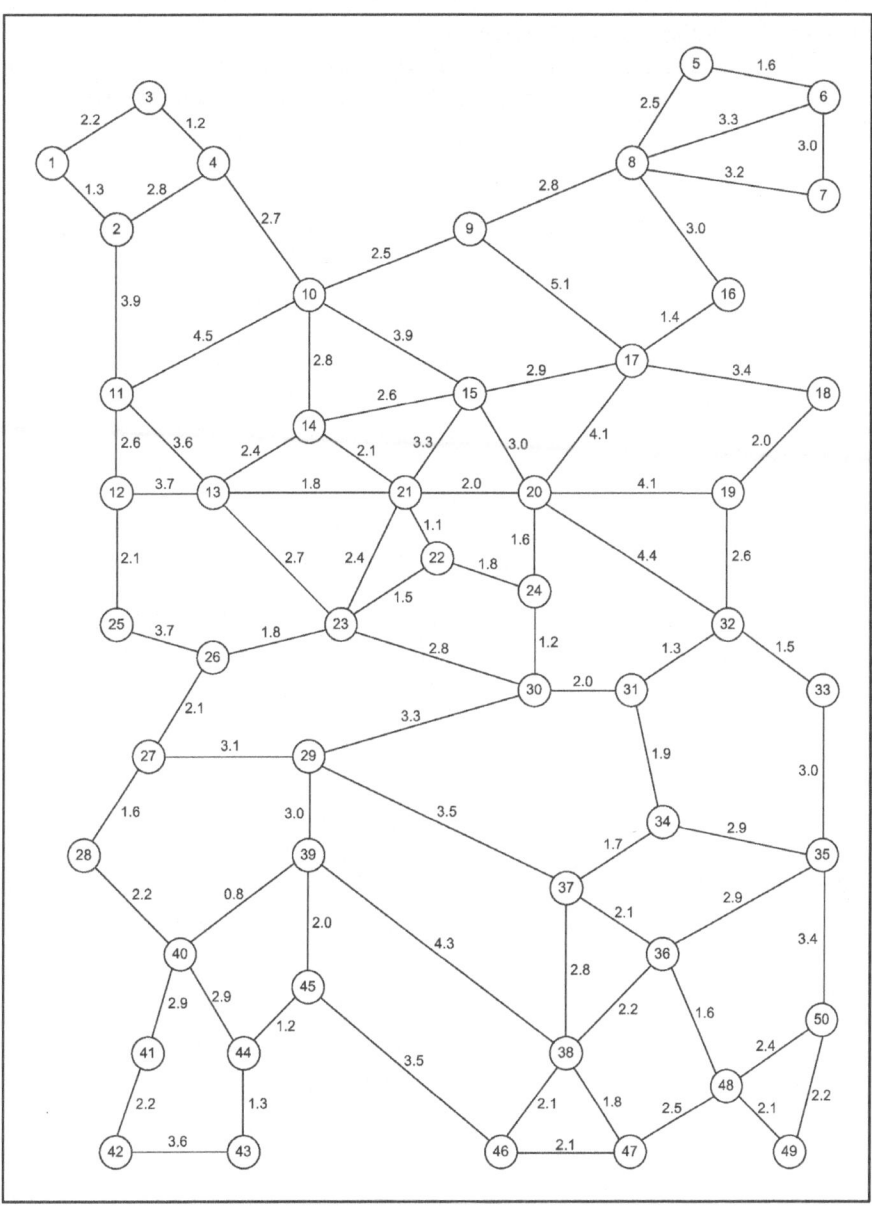

Fig. 2.11. Schematic map with 50 locations (distances in 10 kilometer units, costs in €5,000 per kilometer)

Problem 2.1. Reliable Cable Connections
Since the high capacity cable is expensive, GTC wants to know the minimum length of the cable needed to connect the 50 selected locations from Figure 2.11 in accordance to its contract.

(a) Calculate the minimum length of cable needed to connect all locations in Figure 2.11. Also give the list of connections that are used for the cabling.
(b) GTC regrets that the locations 9 and 17 are not directly connected in the solution of part (a). There can be taken two ways to get 9 and 17 connected — adding a direct cable between 9 and 17 to the solution of Problem 2.1(a) and deleting a most expensive connection in the cycle thus created, or repeating the calculations of Problem 2.1(a) with the extra restriction that the connection 9 – 17 has to be in the new solution. Compare both methods. Will both methods always give the same result? Explain your answer.
(c) The cable system designed in part (a) is not very reliable, in the sense that there is only one connection between any pair of locations. Why is this so?
Determine, by inspection, a most vulnerable connection in your solution to part (a), in the sense that if the cable on this connection breaks down, the most number of pairs of locations will not be able communicate anymore.
(d) Design a reliable cable connection, in the sense that if the cable between any two locations breaks down, there is still a connection (although possibly indirect) between these locations. Is your design the cheapest possible?

Problem 2.2. What Happens If
In another country, GTC can obtain the rights to construct a main cable network that will connect the 46 major towns of this country. The country is very mountainous and a number of large lakes and rivers makes the construction of the cable network very delicate for the environment. These regions are called "vulnerable" and no industrial activity is normally allowed in these regions. Therefore, if GTC lays cable in vulnerable regions, it has to pay the government a certain amount of money, called "environmental price" for environmental restoration activities. In Figure 2.12 the 46 cities are schematically depicted. Attached to the regions (connections between the cities) are the costs of laying the cable (first number) plus the environmental price GTC has to pay to the government (second number).

(a) Is it possible to lay the cables in such a way that each pair of cities are connected and the total cost does not exceed €9,500,000?
(b) Since the cabling project is important for the country, the government is open to negotiating with GTC on regions where the environmental price could be lowered. In case the answer to part (a) is "no", then choose a number of regions on which GTC should start negotiations with the government.
GTC realizes that the government is willing to lower the prices by a maximum of 15% in five regions. Is there a cable system that satisfies the budget of the company and the margins of the government? At what percentage reduction would such a project just be feasible?

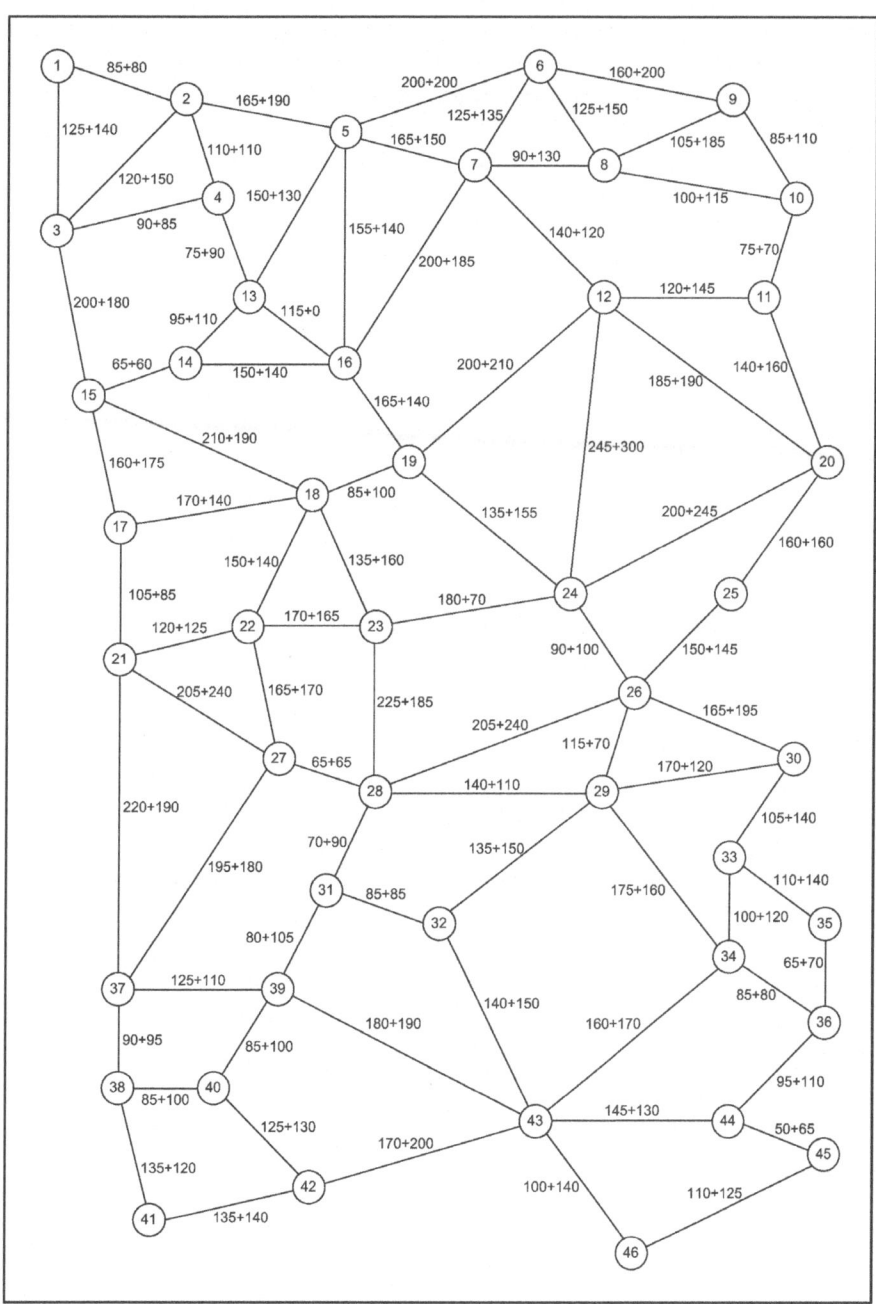

Fig. 2.12. Schematic road map with 46 locations (connection cost = cable cost + environmental cost, costs in €1,000)

(c) Actually, the price of traversing the region between the cities 5 and 13 could not be determined with the same accuracy as the other prices. How much can the price of intersecting the region between 5 and 13 change before a different cable network is more profitable for GTC?

(d) After a consultation round with all people and organizations involved, it is decided that the links 6 – 9 and 15 – 17 have to be included in the new cable network, whereas the links 24 – 26 and 31 – 39 will not be included. Determine a minimal cost solution, without the reductions.

The government was willing to lower the price for the edges 6 – 9 and 15 – 17 by 10% on top of the negotiations from part (b). Was this 10% enough for the project to be financially feasible?

The following problem is based on the network of Figure 2.12. The links 6 – 9 and 15 – 17 have to be included in the cable network, while the links 24 – 26 and 31 – 39 are not in the cable network. The subsidies from the government are not taken into account.

Problem 2.3. Adding Locations to Networks

In the region between the nodes 22, 23, 27, and 28 of Figure 2.12, there is a small town where recently the building activities of a new university are started. The town is expected to expand by at least 10,000 people. The whole building process is expected to take another four years. It is decided that this town will be connected with the main cable system after two years from now. The town is located 20 kilometers from city 22, 28 kilometers from city 23, 20 kilometers from 27, and 25 kilometers from 28.

In order to connect the new university town to the main cable network, GTC considers the following two scenarios:

1. The university town becomes part of the main cable network from the start of the building activities.
2. The town is connected to the main cable network after two years.

In the first scenario the cables have to be maintained for the first two years, which incurs extra costs. In the second scenario all other towns need to be able to communicate during the first two years without the benefit of the links through the university town.

The cost of one kilometer cable is now €5,000, after two years it is expected to drop to €4,000. On the other hand, the price the government asks for recovering the damage is expected to rise as follows: At the moment the prices are €130,000 (for the region between the town and city 22), €110,000 (for the region between the town and city 23), €135,000 (for the region between the town and city 27), and €95,000 (for the region between the town and city 28). After two years, the prices are €145,000, €140,000, €145,000, and €120,000, respectively.

In case of the second scenario, the company would incur an additional cost of €100,000, primarily because all construction equipment must be shipped into the region after two years.

What would be the total costs of maintaining the cables during the first two years in order that the first scenario is more profitable?

Problem 2.4. Expanding Networks
Figure 2.13 is a schematic reproduction of the map of 48 locations labeled 1, ..., 48 in a certain region. For these 48 locations a cable network has to be designed that connects each pair of locations. The numbers attached to the connections denote the costs of laying cable there. However, there are two areas where a cable network was constructed recently: these are $10 - 9 - 5 - 6 - 7$ and $34 - 33 - 28 - 29$ (see the thick lines in Figure 2.13).

(a) Determine a least cost cabling solution for connecting these locations.
(b) The solution of part (a) shows that all connections in location 22 are used in the cable network. This makes this location quite vulnerable. How much worse is a cabling solution if no more than 3 cable connection end at location 22?

Problem 2.5. Grouping Machines for Efficiency
One of the production facilities of GTC has 19 machines and produces 22 products. The facility contains three production halls, and the machines should be placed in these three halls in such a way that similar products (these are products that, to a large extent, need the same machines) are manufactured in the same hall as much as possible. Table 2.1 contains the Machine-Product Incidence (MPI) matrix. The matrix consists of 0's and 1's (the 0's are not shown in the table). An entry of 1 in position (i, j) means that machine i can manufacture product j, while an entry of 0 in that position means that this is not possible. So the similarity between two products is reflected by the similarity of the corresponding columns in the MPI matrix — the more common 0's and 1's, the more similar the corresponding products are.

(a) Determine a grouping of the machines in the three halls using a minimum spanning tree. Take the ratio of the number of products that can be manufactured on only machine M1 to the number of products that can be manufactured on at least one of the two machines M1 and M2 as the "distance" between the two machines M1 and M2. Here, what do you mean by an optimal solution?
(b) What will the grouping be if not more than 10 machines can be placed in any of the halls. Repeat your calculations when the maximum number of machines per hall is 8, 9, 11, ... machines per hall. Compare the results you obtain.
(c) Another production facility of GTC company has a similar problem. The MPI matrix for this facility is depicted in Table 2.2. In this facility, there are 20 machines and 26 products. Answer the questions in parts (a) and (b) for this facility.

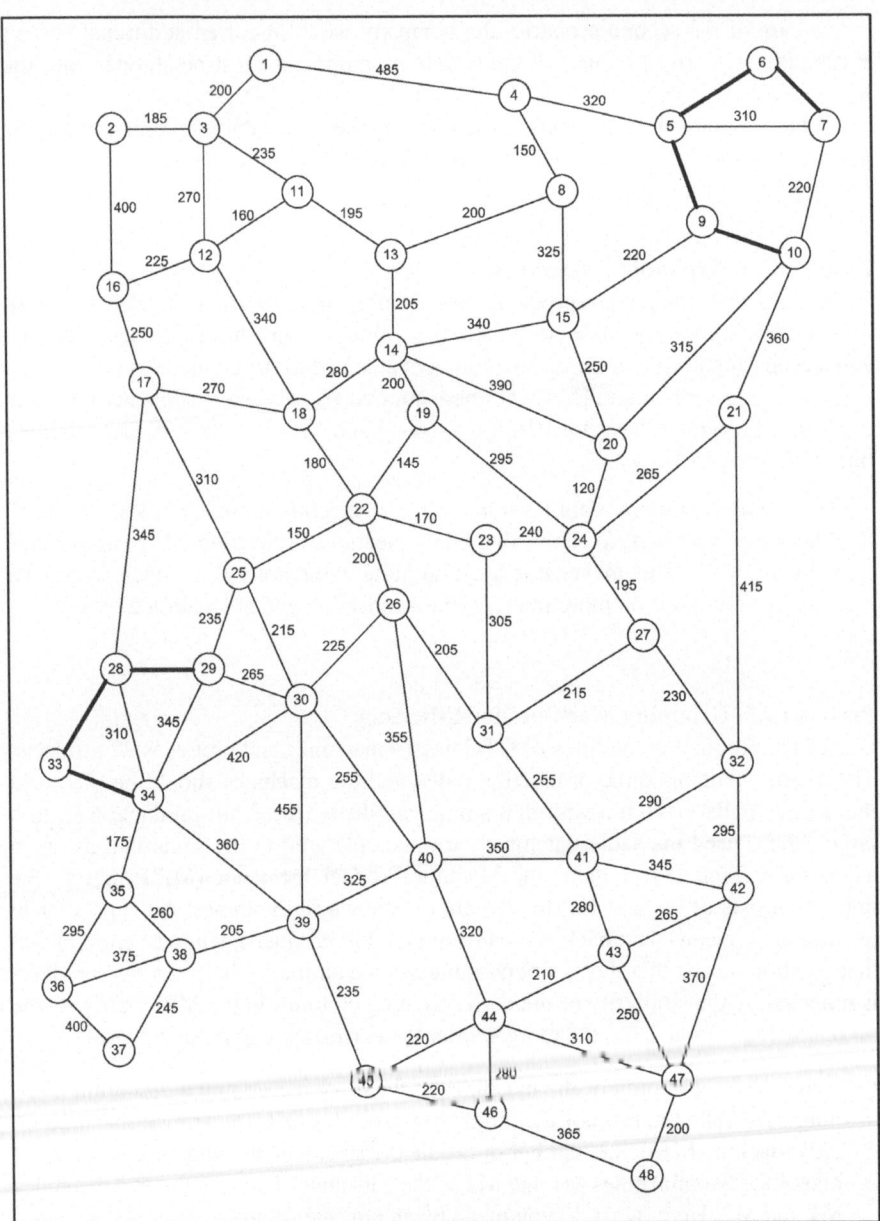

Fig. 2.13. Schematic road map with 48 locations (bold lines refer to existing cables, costs in €1,000)

Table 2.1. Machine-Product Incidence Matrix for Problem 2.5(a,b) (row labels: machines, column labels: products)

	1	2	3	4	5	6	7	8	9	10	11	12	13	14	15	16	17	18	19	20	21	22
1	1	1		1	1			1				1				1			1			
2	1			1				1				1				1			1			
3		1			1			1				1				1				1		1
4	1	1		1	1			1				1				1			1			
5	1	1				1			1			1			1							
6	1	1		1	1			1				1				1			1			
7	1			1					1					1			1			1		1
8			1					1			1				1	1		1				
9		1	1		1	1								1				1				
10	1	1						1			1				1			1				
11		1		1	1				1					1					1			1
12		1		1			1	1			1				1							1
13			1					1				1				1				1		1
14	1	1		1			1	1	1									1	1			
15	1	1		1		1					1			1				1				
16	1	1		1								1			1			1				1
17		1		1	1				1					1					1			
18			1			1					1				1		1		1			
19	1			1				1						1					1			1

Table 2.2. Machine-Product Incidence Matrix for Problem 2.5(c) (row labels: machines, column labels: products)

	1	2	3	4	5	6	7	8	9	10	11	12	13	14	15	16	17	18	19	20	21	22	23	24	25	26
1	1		1	1		1			1				1	1					1					1		
2	1			1			1	1					1		1			1			1			1		1
3	1	1		1		1					1	1				1			1			1				
4	1				1						1				1											1
5		1			1	1	1	1					1				1									
6		1				1						1	1			1		1					1	1	1	
7	1	1		1						1		1		1				1			1		1			
8		1															1							1	1	
9					1	1						1		1				1					1		1	
10	1		1		1					1								1					1			
11							1	1	1						1	1			1							1
12	1	1	1								1			1				1					1			
13						1	1					1	1			1	1				1					
14			1	1	1							1		1			1	1	1	1						
15	1	1						1							1								1			1
16	1		1		1					1				1	1				1	1			1			
17		1		1					1					1			1	1								
18	1					1					1				1								1			
19		1									1	1				1	1	1	1						1	
20		1	1			1	1											1				1				

Problem 2.6. Designing a Radio Telescope

The astronomical society ASTRO wants to build a large radio telescope. The telescope will consist of thirteen sensors laid out in a four armed spiral in the countryside. Figure 2.14 shows the location of the sensors marked A, B, ..., M. Each of these sensors continuously collect vast amounts of data, and these data need to be sent to a central processing station located at A for analysis. GTC has been asked to provide the physical connection to make this data transfer possible. The cable connections that ASTRO requires need to be reliable, and so the cost of laying cable is quite high, about €10 per meter of cable laid.

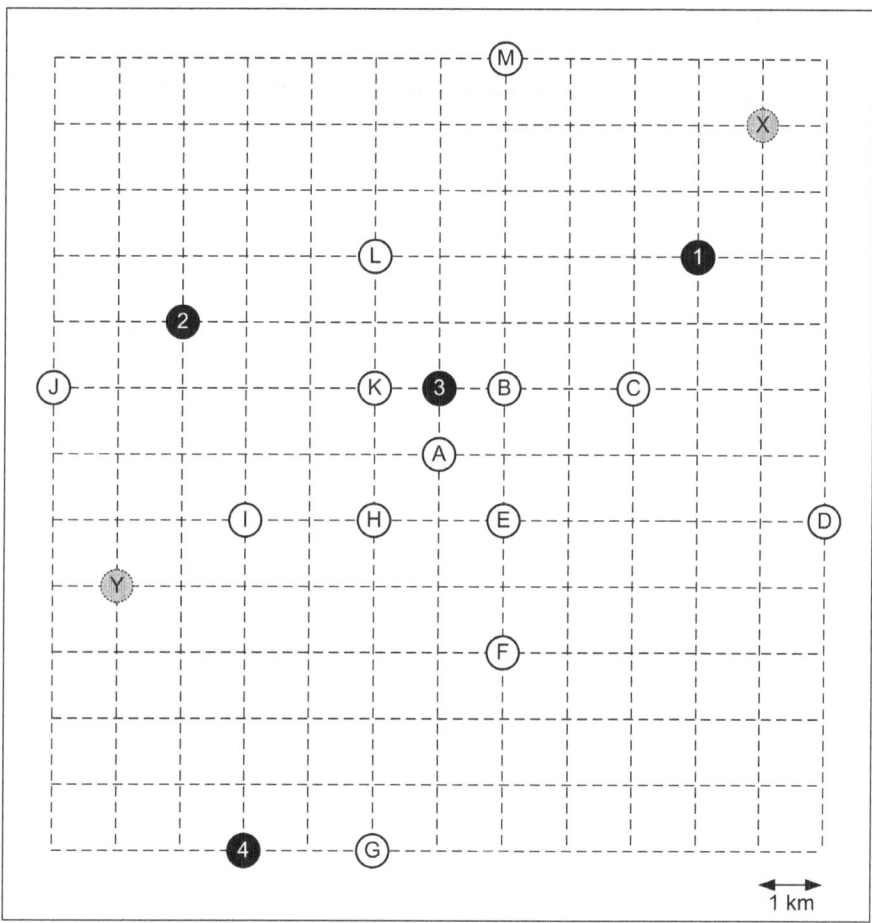

Fig. 2.14. Schematic map of the telescope sensors, the access points, and the cities

(a) What is GTC's minimum investment in cable laying in order to complete the project?

GTC knows that the JOM telecommunication company already has cable laid out in the area, and is wondering whether it can make use of JOM cables to reduce ASTRO's cabling costs. JOM has indicated that it can provide four access points in the area (marked 1, 2, 3, and 4 in Figure 2.14), and that these access points are connected to each other. GTC is considering a deal with JOM whereby it will pay JOM some money to use JOM's network via the access points to transmit data.

(b) What is the maximum amount that GTC should be willing to pay JOM?

Under the previous scheme, JOM has agreed to charge GTC €23,500 for using its access points. Executives at JOM inform GTC of a second scheme. They say that under the new scheme, GTC could use one or more of their access points in the area by paying €7,500 per access point used.

(c) Is this new scheme more attractive to GTC than the previous one?

Two neighboring villages X and Y want to make use of the connections to link themselves up to one another and to the JOM network. They are willing to pay GTC €25,000 each to connect them to the network.

(d) Is it profitable for GTC to connect the villages to the network? Which scheme of JOM should GTC consider if they plan to connect the two villages to the JOM network?

3

Network Flows

3.1 Introduction

The schematic diagram of the telecommunication network of GTC in a particular region is shown in Figure 3.1. Each node in the network represents a major network

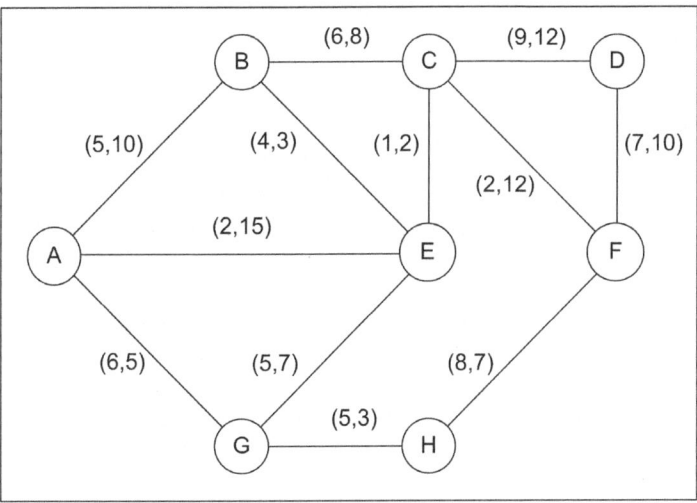

Fig. 3.1. Existing telecommunication network

hub, while each edge in the network represents a link between two such hubs. In the figure, there is a vector containing two numbers along each edge. The first entry of the vector indicates the cost of sending one GBps of data along that edge (in €100 units), while the second entry indicates the bandwidth available along that edge (in 100GBps units). GTC is considering a request to provide a bandwidth of 600GBps between the points A and H in the network and is wondering what the minimum cost of providing this bandwidth would be.

G. Sierksma and D. Ghosh, *Networks in Action: Text and Computer Exercises in Network Optimization*, International Series in Operations Research & Management Science 140, DOI 10.1007/978-1-4419-5513-5_6, © Springer Science + Business Media, LLC 2010

This problem is a typical example of a minimum cost flow problem. The solution to the problem consists of a collection of edges in the network, which can be combined to form paths, and the amount of flow along each of the edges. The sum of the costs of this collection of edges, each weighted by the amount of flow it carries, is the minimum cost that has to be incurred to route the required amount of flow.

In GTC's case, GTC has to come up with a set of paths from A to H, such that the sum of the capacities of the paths is at least 600GBps. However, if more than one path includes the same edge, then the sum of the flows along the paths cannot exceed the capacity of the common edge. For example, a solution including paths A – G – H and A – E – G – H has a total capacity of 300GBps although the individual capacities of both the paths are 300GBps each, since both the paths include the common edge G – H. See that a feasible solution to GTC's problem is given by the collection of arcs A – B, A – E, A – G, B – C, E – C, C – F, F – H, and G – H, which constitutes the paths A – G – H, A – E – C – F – H, and A – B – C – F – H. In the solution, the path A – G – H provides a bandwidth of 300GBps bandwidth, the path A – E – C – F – H provides 200GBps, and the path A – B – C – F – H provides the remaining 100GBps. The total cost of this solution is €800,000.

Formally stated, in a *minimum cost flow problem*, one is given a connected network $N = (V, E, C, K)$ where $C = (c_e)$ is a vector of costs of each edge $e \in E$, and $K = (k_e)$ is a vector of capacities along each edge $e \in E$. One is also given a source node $s \in V$, a destination node $t \in V$, and an amount B of flow that has to be sent from s to t through the network. A solution to the problem is a collection of edges $E' \subseteq E$ and flows f_e along the edges $e \in E'$, such that the collection corresponds to a set of paths from s to t whose combined capacity is enough to send B units of flow from s to t. The cost of the solution is the sum $\sum_{e \in E'} c_e f_e$. One is required to find a solution having the minimum possible cost.

A special case of the minimum cost flow problem is one where the amount of flow to be sent from the source node to the destination node does not exceed the capacities of each of the arcs, i.e., $B \leq \min_e \{k_e\}$. In this case, all the flow would be sent through one path only, and that path is the cheapest path from the source node to the destination node through the network. This happens in particular when the amount of flow to be transmitted satisfies $B = 1$, and all edge capacities are infinite. Notice that the problem then is identical to the shortest path problem described in Chapter 1. Therefore the shortest path problem can be considered as a special case of the minimum cost flow problem.

Next consider another problem. A trucking company that has to move truckloads of cables from point to point along a road network. The road network can be represented by Figure 3.2, in which each of the arcs correspond to a one-way road segment. The two numbers along each arc in the figure correspond to the cost of transporting one truckload of cable along that road segment and the capacity (in truckloads/hr units) of the road segment. We need to find out the maximum rate at which cable can be transported from node A to node F in the network.

An inspection of the road network shows that it is possible to tansport a maximum of 80 truckloads of cable per hour. Each hour, 20 truckloads can pass through the segment A – B – C – F, 10 truckloads through A – B – E – F, 30 truckloads through

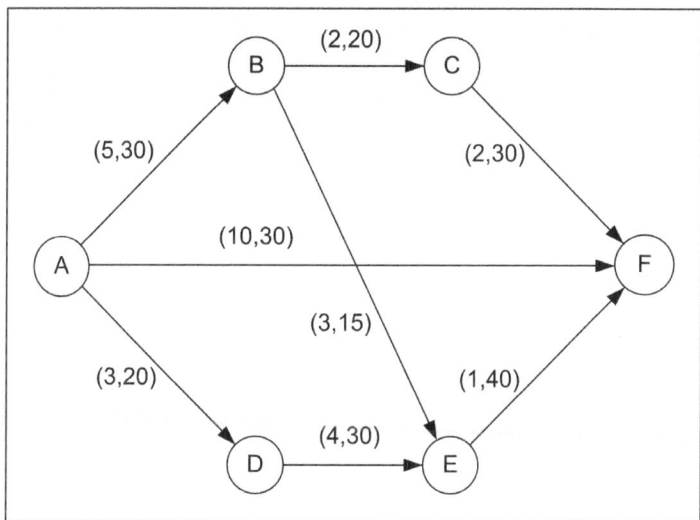

Fig. 3.2. A road transport network

A – F, and a further 20 truckloads through A – D – E – F. Notice that the costs of transportation of cables are superfluous in this situation, because the situation does not restrict the company's transportation plan through a budget. This type of problem is commonly called the maximum network flow problem.

Formally stated, in a *maximum network flow problem*, one is given a network $N = (V, E, K)$ where $K = (k_e)$ is a vector denoting the capacities of each $e \in E$, a source node $s \in V$, and a destination node $t \in V$. One is required to find the maximum amount of flow that can be routed through the network N from s to t without violating any capacity restriction.

3.2 Applications

Network flow problems arise very frequently in real-life applications. In the remainder of the section, we outline some illustrative situations where these problems can be solved to aid decisions.

3.2.1 Production planning

Production processes that involve more than one operations can often be represented naturally as networks. This representation allows managers to answer various questions about the processes by solving network flow problems.

Consider for example, a production process involving two steps, a preprocessing step and a finishing step. Assume that the preprocessing step can be done at two sites, A and B, and the finishing step can be done at two other sites, C and D. The output of preprocessing at site A can be finished at either C or D, while the output of preprocessing at B can only be finished in D. Site A can preprocess up to 200 tonnes of raw material each month, while site B can preprocess up to 150 tonnes. Site C can finish up to 50 tonnes of material each month, while site D can finish up to 200 tonnes. The cost of operation at sites A, B, C, and D are €40, €20, €15, and €30 per tonne, respectively. Any amount of material can be transported to each of the sites, and from C and D to the finished materials site. The cost of such transportation is €10 per tonne.

This production process can be represented by the network in Figure 3.3. In the

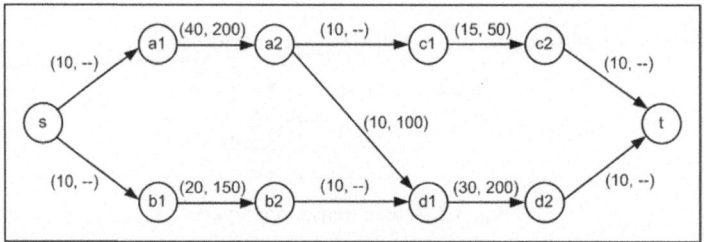

Fig. 3.3. A network representation of the production process

figure, the node s represents the source of the raw material and the node t represents the finished materials shed. Site A is represented by the pair of nodes $a1$ and $a2$ and the arc connecting them. A unit of flow through $a1 \to a2$ represents a unit of material being processed at site A. Sites B, C, and D are represented in a similar manner.

If a manager wants to supply a demand of, say, 200 tonnes each month, and wants to find the cheapest way to produce it, then she can solve a minimum cost flow problem on the network in Figure 3.3 after adding an inflow of 200 tonnes at s and an outflow of 200 tonnes at t. If the manager wants to find out the maximum production capacity of the production process, then she can solve a maximum flow problem from s to t on the same network.

3.2.2 Tourist reservation system

Consider a tourist route that starts at point A, goes through point B, and ends at point C. All of A, B, and C are tourist destinations, and tourists would like to book passage on a tourist bus to either go from A to B, or from A to C, or from B to C. A tour operator wants to decide how many tourists to pick up at points A and B in order to maximize fare collection. Suppose that the tourist bus can carry 80 tourists at any point of time. On a particular day, 50 tourists want to go from A to B, 40 tourists want to go from A to C, and 60 tourists want to go from B to C. The fare from A to B is €100, from A to C is €170, and from B to C is €120.

The tour operator's problem can be modeled using the network shown in Figure 3.4. The arcs in the network have costs and capacities as usual. Each unit of flow in the network represents the fate of one tourist in the reservation system. In

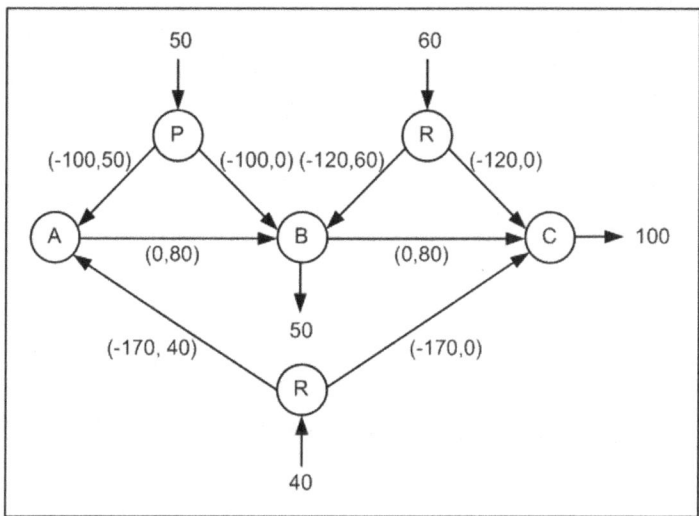

Fig. 3.4. A network representation of the tour operator's problem

the figure, nodes A, B, and C correspond the three points of the tourist route. The inflow at node P represents the demand for passage from A to B, the inflow at node Q represents the demand for passage from A to C, the inflow at node R represents the demand for passage from B to C. The inflow at node P can either be routed along the path P → A → B, or through P → B to reach node B. The former path represents tourists who are sold tickets on the bus, and the latter represent tourists who want to go from A to B but do not succeed to obtain tickets on the bus. Notice that not more than 80 units can flow through P → A → B due to the capacity restriction on A → B. Similar routings are possible for inflows at nodes Q and R. The outflows at B and C ensure that a tourist who obtains a ticket is taken to precisely that destination that she or he wants to reach.

Solving a minimum cost flow problem on this network, given the inflows and the outflows, provides the tour operator with the decision on how many tourists of each category to book on the bus. The flow on P → A gives the number of tourists to book who want to go from point A to point B. Similarly, the flows on Q → A and R → B gives the numbers of tourists to book who want to go from point A to point C, and from point B to point C, respectively. The objective function value multiplied by −1 gives the maximum revenue from ticket sales that the operator can expect.

3.2.3 Staff allocation

Consider the network of highways between six cities labeled A, B, ..., F shown in Figure 3.5. The edges in the network represent highways and the numbers next to the edges represent the number of lanes in the highway. According to intelligence reports, there is a possibility of criminals transporting contraband from A to F through this network. In order to intercept the contraband, the police department wants to

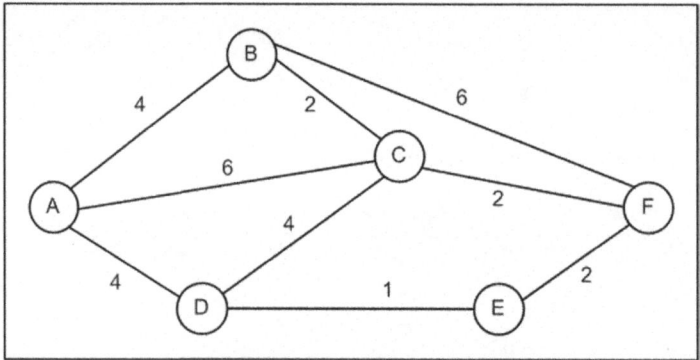

Fig. 3.5. A network of highways connecting six cities

position officers along the network to check vehicles. A police officer can check vehicles coming down one lane at a time. Therefore, since the highway segment between A and B is a four lane highway, the department has to assign four police officers to check for contraband along it. The police department wants to find out the minimum number of police officers that it requires to check for contraband.

Notice that contraband can be transported using any combination of lanes in the highway network. So in order to find the minimum number of police officers that need to be assigned on the job, one needs to find out the maximum number of lane combinations possible to go from A to F. This can be found by finding out the maximum flow that can be sent from A to F through the network, assuming each unit of flow to be a vehicle being sent through a lane combination denoting a path from A to F. The maximum number of units that can be sent from A to F through the network therefore is also the minimum number of officers who need to be assigned to ensure that the contraband is intercepted if it is sent through the network.

The decision problem of finding out on which highway segments to assign police officers can also be solved through a related problem called a minimum cut problem. The maximum flow problem can be solved using a linear program (see the next section). From the celebrated maximum flow - minimum cut theorem, it can be seen that the shadow prices for the capacity constraints in this formulation assume values of either 1 or 0. It turns out that assigning police officers to those highway segments whose capacities have a shadow price of 1 would ensure that the contraband would be intercepted, while assigning the minimum number of officers.[1]

---

[1] For a more detailed treatment on the maximum flow - minimum cut theorem, see C.H. Papadimitriou and K. Steiglitz, Combinatorial Optimization: Algorithms and Complexity, Prentice-Hall, Inc., USA, pp. 117–120.

3.3 Linear Programming Formulations

Both the minimum cost flow problem and the maximum network flow problem can be modeled effectively as linear programs. In this section we describe these formulations. In formulations of network flow problems, whenever a flow is possible from a node i to a node j in the network, a decision variable x_{ij} is defined to denote the amount of flow.

Since edges in networks allow flows in both directions, each edge $i - j$ between two nodes i and j in the network is replaced by a pair of arcs, $i \rightarrow j$ from i to j, and $j \rightarrow i$ from j to i. If the edge $i - j$ had a cost c_{ij} associated with it, both the arcs $i \rightarrow j$ and $j \rightarrow i$ have the same cost c_{ij} associated with them. The capacities of the two arcs depend on the situation being modeled. Let us consider an example of bandwidth allocation, when there is a flow of x_{ij} along arc $i \rightarrow j$ and a flow of x_{ji} on $j \rightarrow i$. Then the implication is a residual flow of $|x_{ij} - x_{ji}|$ along edge $i - j$. In such a case, the individual capacities on the arcs are not important, but flows along these arcs need to obey the constraint

$$-k_{ij} \leq x_{ij} - x_{ji} \leq k_{ij}, \tag{3.1}$$

where k_{ij} is the capacity of edge $i - j$.

In certain other situations however, the allocation of capacities is handled differently. For instance in a road network with roads which allow traffic in both directions, the capacity along a road segment denotes the number of cars that can travel on that segment during a period. If we replace such an edge with two arcs, then we need to ensure that the sum of the flows along the two arcs does not exceed the capacity of the edge. In other words, in this situation, if an edge $i - j$ has a capacity k_{ij}, and the edge is replaced with two arcs $i \rightarrow j$ and $j \rightarrow i$, then each of the arcs has a capacity of k_{ij}. In addition, the flows x_{ij} and x_{ji} along the two arcs must obey the constraint

$$x_{ij} + x_{ji} \leq k_{ij}. \tag{3.2}$$

Once we have converted edges to arcs in a network, we have a directed network. Let A denote the set of arcs in the directed network that we generate. In a *minimum cost flow problem*, the objective is to minimize the cost of transmitting the required amount of flow from the source node to the destination node. So given that the cost of transmitting one unit of flow from node i to node j is c_{ij}, the objective in a minimum cost flow problem is to

$$\text{Minimize} \sum_{i \rightarrow j \in A} c_{ij} x_{ij}. \tag{3.3}$$

There are two sets of constraints that define feasible flows along the network. In the first set of constraints, the net outflow at the source node equals the net inflow at the destination node, and both are equal to the amount of flow that has to be transmitted. For all other nodes in the network, the net flow, i.e., the difference between the outflow and the inflow at the node must be zero. This leads to the following constraints, for each $k \in V$,

$$\sum_{i:i\to k\in A} x_{ik} - \sum_{j:k\to j\in A} x_{kj} = \begin{cases} -B \text{ if } k \text{ is the source node} \\ B \text{ if } k \text{ is the destination node} \\ 0 \text{ otherwise;} \end{cases} \quad (3.4)$$

where B is the total amount of flow to be transmitted.

These constraints are called the *flow balance constraints*. Notice that in our formulation the left hand side of the constraint corresponds to the net inflow at any node. We could also rewrite these constraints to represent the net outflow at any node. The flow balance constraints in such cases would look like

$$\sum_{i:k\to i\in A} x_{ki} - \sum_{j:j\to k\in A} x_{jk} = \begin{cases} B \text{ if } k \text{ is the source node} \\ -B \text{ if } k \text{ is the destination node} \\ 0 \text{ otherwise.} \end{cases} \quad (3.5)$$

It is advisable to restrict oneself to one of these conventions while formulating network flow problems.

The second set of constraints in the formulation is concerned with the capacities of flows along the arcs. For arcs that have individual capacities specified, we could use constraints enforcing these capacities. For example, if a link $i \to j$ has a capacity of k_{ij} then it is expressed in the formulation by the constraint

$$x_{ij} \le k_{ij}. \quad (3.6)$$

However, if no such individual capacities are available, for example when the arcs are a result of creating a directed network from an undirected network, one may use constraints of the form (3.1) or (3.2) described earlier.

This formulation to obtain an optimal, i.e., minimum cost routing of a flow of magnitude B from a source node s to a destination node t through a network $N = (V, E, w)$ where the edge set E has been replaced with an appropriate arc set A is shown in its entirety in Figure 3.6.

Figure 3.8 shows the formulation of the example of the minimum cost flow problem described in Section 3.1. The network for this problem, shown in Figure 3.1, is an undirected network, and the first step is to create a directed network from it. This is done by simply replacing each edge $i - j$ in the network with a pair of arcs $i \to j$ and $j \to i$. The directed network is shown in Figure 3.7 in which the numbers next to each pair of arcs denote the costs of sending flows along the arcs.

Maximum network flow problems can also be solved using linear programming. As in other network flow formulations, the decision variables that are used are x_{ij}'s, one for each arc $i \to j$ in the network, denoting the amount of flow that is to be sent along that arc. The objective is to maximize the flow from the source node (or the flow to the destination node). This flow is normally represented in a novel way using a "back arc". A back-arc is an artificially added arc from the destination node to the source node. This arc has infinite capacity. Thus in principle, any amount of flow sent from the source node to the destination node can return via the back-arc to the source node. Given the problem described on the network in Figure 3.2, the network in Figure 3.9 shows the modified network with the back-arc. The back arc is represented in the figure with broken lines.

Minimize

$$z = \sum_{i \rightarrow j \in A} c_{ij} x_{ij}$$

Subject to

$$\sum_{i:k \rightarrow i \in A} x_{ki} - \sum_{j:j \rightarrow k \in A} x_{jk} = \begin{cases} B & \text{if } k \text{ is the source node} \\ -B & \text{if } k \text{ is the destination node} \\ 0 & \text{otherwise.} \end{cases} \quad \text{for each } k \in V$$

$$x_{ij} \leq k_{ij} \qquad \text{for each } i \rightarrow j \in A$$

$$x_{ij} \geq 0 \qquad \text{for each } i \rightarrow j \in A$$

Fig. 3.6. Linear programming formulation of the minimum cost flow problem

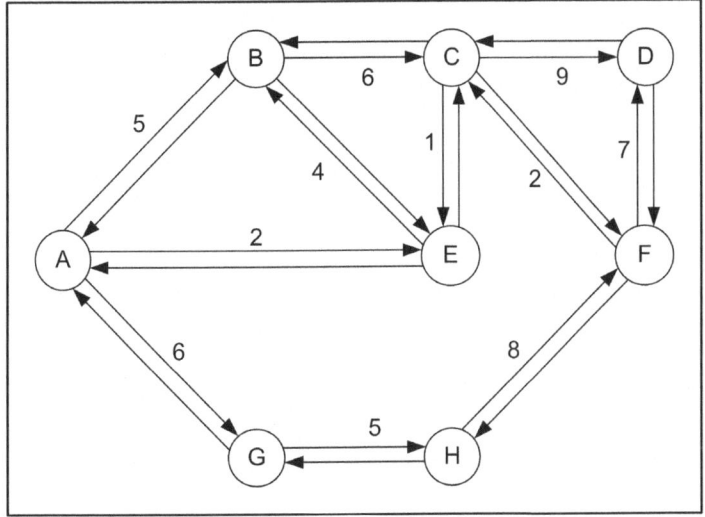

Fig. 3.7. Directed version of network in Figure 3.1

The linear program then aims to maximize the flow along the back arc while maintaining the flow balance constraints at the nodes, and obeying capacity constraints on the arcs. Mathematically expressed, therefore, the objective is to

$$\text{Maximize } x_{back\text{-}arc} \qquad (3.7)$$

where $x_{back\text{-}arc}$ is a decision variable representing the flow along the back-arc.

Minimize

$$z = 5x_{AB} + 5x_{BA} + 2x_{AE} + 2x_{EA} + 6x_{AG} + 6x_{GA} + 6x_{BC} + 6x_{CB} + 4x_{BE} + 4x_{EB} + 9x_{CD} + 9x_{DC} +$$
$$x_{CE} + x_{EC} + 2x_{CF} + 2x_{FC} + 7x_{DF} + 7x_{FD} + 5x_{EG} + 5x_{GE} + 8x_{FH} + 8x_{HF} + 5x_{GH} + 5x_{HG}$$

Subject to

$$\begin{array}{rll}
(x_{BA} + x_{EA} + x_{GA}) - (x_{AB} + x_{AE} + x_{AG}) = & -600 & \text{(Condition (3.4) at A)} \\
(x_{GH} + x_{FH}) - (x_{HG} + x_{HF}) = & 600 & \text{(Condition (3.4) at H)} \\
(x_{AB} + x_{CB} + x_{EB}) - (x_{BA} + x_{BC} + x_{BE}) = & 0 & \text{(Condition (3.4) at B)} \\
(x_{BC} + x_{DC} + x_{EC} + x_{FC}) - (x_{CB} + x_{CD} + x_{CE} + x_{CF}) = & 0 & \text{(Condition (3.4) at C)} \\
(x_{CD} + x_{FD}) - (x_{DC} + x_{DF}) = & 0 & \text{(Condition (3.4) at D)}
\end{array}$$

There are three more similar constraints for nodes E, F, and G.

$$\begin{array}{rll}
x_{AB} - x_{BA} \leq & 10 & \text{(Capacity on A – B)} \\
x_{AB} - x_{BA} \geq & -10 & \text{(Capacity on A – B)} \\
x_{AE} - x_{EA} \leq & 15 & \text{(Capacity on A – E)} \\
x_{AE} - x_{EA} \geq & -15 & \text{(Capacity on A – E)}
\end{array}$$

There are twenty more capacity constraints on all other edges.

$$x_{AB}, x_{BA}, x_{AE}, x_{EA}, x_{AG}, x_{GA}, x_{BC}, x_{CB}, x_{BE}, x_{EB}, x_{CD}, x_{DC},$$
$$x_{CE}, x_{EC}, x_{CF}, x_{FC}, x_{DF}, x_{FD}, x_{EG}, x_{GE}, x_{FH}, x_{HF}, x_{GH}, x_{HG} \geq 0 \quad \text{(Nonnegative flows)}$$

Fig. 3.8. Formulation of the minimum cost flow problem for the minimum cost flow problem in Section 3.1

There are two sets of constraints in a linear programming formulation to solve maximum network flow problems. The first set of constraints is the set of flow balance constraints. Since there are no external inflows and outflows, the flow balance constraint for any node k is

$$\sum_{i:i \to k \in A} x_{ik} - \sum_{j:k \to j \in A} x_{kj} = 0. \tag{3.8}$$

As in the case of minimum cost flow problems, this constraint can also be written as

$$\sum_{j:k \to j \in A} x_{kj} - \sum_{i:i \to k \in A} x_{ik} = 0. \tag{3.9}$$

The second set of constraints is the set of capacity constraints. For each arc $i \to j$ with capacity k_{ij}, the constraint is

$$x_{ij} \leq k_{ij}. \tag{3.10}$$

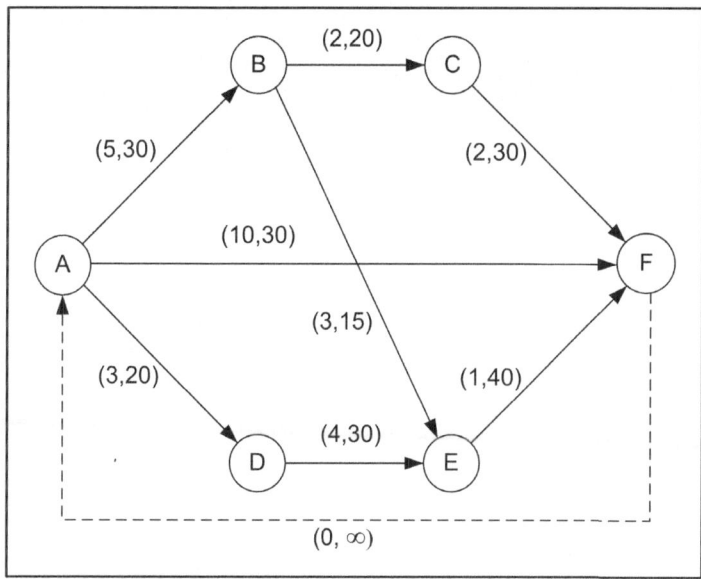

Fig. 3.9. The network in Figure 3.2 with back-arc

Note that since the back-arc has infinite capacity, it is not necessary to include a capacity constraint for the back-arc.

Figure 3.10 shows the full formulation of the maximum flow problem on a network $N = (V, E, K)$ where the edge set E has been appropriately replaced with an arc set A, and where the source node is s ($\in V$) and the destination node is t ($\in V$). A back-arc $t \rightarrow s$ of infinite capacity is assumed to have been included in A. The formulation that results from the implementation of this model on the maximum network flow problem situation described in Section 3.1 is shown in Figure 3.11.

Minimize

$$z = x_{ts}$$

Subject to

$$\sum_{i:k\rightarrow i\in A} x_{ki} - \sum_{j:j\rightarrow k\in A} x_{jk} = 0 \quad \text{for each } k \in V.$$

$$x_{ij} \leq k_{ij} \quad \text{for each } i \rightarrow j \in A, \text{except } t \rightarrow s$$

$$x_{ij} \geq 0 \quad \text{for each } i \rightarrow j \in A$$

Fig. 3.10. Linear programming formulation of the maximum network flow problem

Minimize

$$z = x_{FA}$$

Subject to

$$(x_{FA}) - (x_{AB} + x_{AD} + x_{AF}) = 0 \quad \text{(Flow balance at A)}$$
$$(x_{AB}) - (x_{BC} + x_{BE}) = 0 \quad \text{(Flow balance at B)}$$

There are four more constraints implementing flow balance at D, E, and F.

$$x_{AB} \leq 30 \quad \text{(Capacity of } A \rightarrow B)$$
$$x_{AD} \leq 20 \quad \text{(Capacity of } A \rightarrow D)$$

There are seven more capacity constraints for the other arcs.

$$x_{AB}, x_{AD}, x_{AF}, x_{BC}, x_{BE}, x_{CF}, x_{DE}, x_{EF}, x_{FA} \geq 0 \quad \text{(Nonnegative flows)}$$

Fig. 3.11. Linear programming formulation for the maximum network flow problem in Section 3.1

3.4 Algorithms for Network Flow Problems

Apart from linear programming, there are other efficient algorithms for solving network flow problems. Among these, the algorithms for solving minimum cost flow problems are too advanced for this text. Hence we discuss only one algorithm for the maximum network flow problem in this section.

3.4.1 Ford-Fulkerson's algorithm

A commonly used algorithm for determining the maximum flow that can be routed through a network from a given source node to a given destination node is described in this section. This algorithm is due to L.R. Ford and D.R. Fulkerson (1956). It tries to find successive paths from the source node to the destination node which can send more flow between them. Since each of the paths thus found increase the total flow possible from the source node to the destination node, they are called *augmenting paths*.

Given a network, the first step is to find out a path from the source node to the destination node. This is an augmenting path. The amount of flow that can be sent through this path is obviously the minimum of the capacities of the arcs along this path. This flow is often referred to as an augmenting flow. Once an augmenting flow has been sent along this path, the algorithm needs to find out whether more any flow can be sent from the source node to the destination node, using other paths in the network. This check is done by forming the residual network. In the residual network, the capacities of the arcs that were not on the augmenting path do not change in any way. The capacities of the arcs along the augmenting path however do change. There are two types of arcs on the augmenting path: the unsaturated ones, for which the amount of flow is less than the arc capacity, and the saturated arcs, for which the flow is equal to the arc capacity. In the residual network, the capacities of the unsaturated arcs are reduced by the augmenting flow value.

The way the algorithm deals with saturated arcs is more interesting. These arcs are removed from the residual network and are replaced with arcs in the opposite direction, with the same capacity as the original saturated arc. This operation seems counter-intuitive, but has an interesting explanation. Let v_s and v_d be the source and destination arcs, and $i \rightarrow j$ be a saturated arc along the augmenting path $v_s \rightsquigarrow p \rightarrow i \rightarrow j \rightarrow q \rightsquigarrow v_d$. The notation $v \rightsquigarrow w$ means that there is a path from node v to node w that includes other non-specified nodes. Let the augmenting flow be of amount f_1. Since $i \rightarrow j$ is saturated by the augmenting flow, no more flow can be sent along the path from i to j. However, if the replacement suggested above is accepted, then there exists an arc $j \rightarrow i$ in the residual network with the capacity of the original $i \rightarrow j$. Assume that there is an augmenting path $v_s \rightsquigarrow k \rightarrow j \rightarrow i \rightarrow l \rightsquigarrow v_d$. Let the augmenting flow on this path be f_2. Clearly $f_2 \leq f_1$. Then both augmenting flows can be realized by sending a flow amounting to $f_1 - f_2$ along $v_s \rightsquigarrow p \rightarrow i \rightarrow j \rightarrow q \rightsquigarrow v_d$, and flows amounting to f_2 along each of $v_s \rightsquigarrow p \rightarrow i \rightarrow l \rightsquigarrow v_d$ and $v_s \rightsquigarrow k \rightarrow j \rightarrow q \rightsquigarrow v_d$.

Once the residual network has been formed after sending one augmenting flow, the algorithm tries to send more augmenting flows along the residual network. It stops when no augmenting flow is possible in the residual network at any stage. The maximum flow that can be sent from the source node to the destination node is the sum of the augmenting flows obtained at each step. For the part of the maximum flow that an individual arc $i \rightarrow j$ carries, we compute the sum of flows along $i \rightarrow j$ in each iteration of the algorithm and subtract from it the sum of flows along $j \rightarrow i$ in each iteration.

As an example to illustrate the working of the algorithm, it is applied here to find the maximum flow from node A to node F in the network shown in Figure 3.2. The residual network at the beginning of each iteration is shown in the left hand side of Figure 3.12, and the augmenting path found in that iteration is shown in the right, with the flow depicted by thicker lines. The numbers along each of the arcs in each of the networks in Figure 3.12 refer to the capacities along the arcs.

The residual network at the beginning of the first iteration is the original network in Figure 3.2. $A \rightarrow B \rightarrow C \rightarrow F$ is an augmenting path in this network, and the augmenting flow is of $\min\{30, 20, 30\} = 20$ units. Arcs $A \rightarrow B$ and $C \rightarrow F$ are unsaturated, and $B \rightarrow C$ is saturated. Therefore the residual network at the beginning of the second iteration has the capacities of $A \rightarrow B$ and $C \rightarrow F$ reduced by 20 units, and the arc $B \rightarrow C$ replaced by arc $C \rightarrow B$ with capacity 20 units.

The steps of the algorithm on the network in Figure 3.2 are described in Table 3.1. The maximum flow that can be sent over the network from node A to node F is

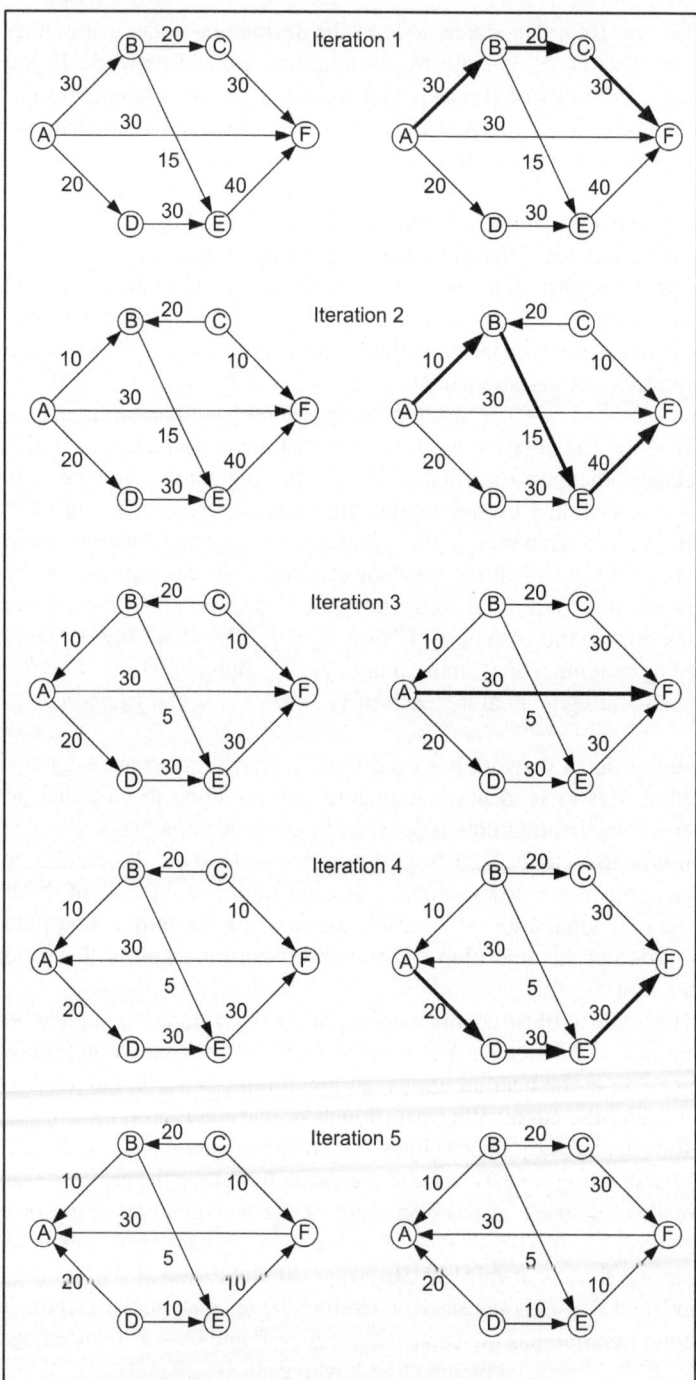

Fig. 3.12. Ford Fulkerson's algorithm in action

Table 3.1. Ford-Fulkerson's algorithm at work

Iteration	Augmenting Path	Augmenting Flow
1	$A \to B \to C \to F$	20
2	$A \to B \to E \to F$	10
3	$A \to F$	30
4	$A \to D \to E \to F$	20
5	none	0

$20 + 10 + 30 + 20 = 80$ units. $A \to B$ carries 30 units of this flow, $A \to F$ carries 30 units, $A \to D$ carries 20 units, $B \to C$ carries 20 units, $B \to E$ carries 10 units, $D \to E$ carries 20 units, $C \to F$ carries 20 units, and $E \to F$ carries 30 units.

3.5 Other Network Flow Problems

3.5.1 The multicommiodity flow problem

Consider a country-wide rail network through which two commodities, wheat and cement, have to be transported. The same rakes are used to transport both commodities. The demand and supply points for both commodities are known in advance, as are the requirements and supply capacities of both commodities at the individual demand and supply points. The rake availabilities and transportation costs are also known along each link in the rail network. In essence therefore, the problem is to transport the two commodities so that the demands for the commodities are met in a way such that the total transportation cost is minimized. The problem does not decompose into two independent minimum cost flow problems since the two commodities share the same rakes. Such problems are known as multicommodity flow problems.

In formal terms, in a *multicommodity flow problem*, one is given a network $N = (V, A, C, K_1, K_2, \ldots, K_m)$ where for each $i = 1, \ldots, m$, K_i is the vector of the costs of sending one unit of flow of type i through each arc of the network. One is also given a set of source nodes for each flow with their supply capacities and a set of destination nodes for each flow with their demands. The objective is to route the flows through the network at minimum cost such that all demands are met.[2]

3.5.2 The reliable network flow problem

Consider a telecommunication network which is used by several branches of a bank to process transactions with a central computer at the main branch. Each bank generates a certain number of transactions per unit time, and the computer at the main

---

[2] For more details on this problem, see R.K. Ahuja, T.L Magnanti, and J.B. Orlin, Network Flows: Theory Algorithms and Applications, Chapter 17: Multicommodity Flows, 1992, Prentice Hall.

branch is capable of handling all transactions. However the network that the bank uses consists of links which are not reliable; one or more links can go down at any point of time. This would cause the network to fail to communicate all transactions from the branches to the central computer. The bank's objective is to find out what proportion of transactions it can expect will go through with 90% probability. As a related problem, it may want to re-design the network in order for it to allow all transactions to go through at any point of time with say, 90% probability. These problems are known as reliable network flow problems.

Formally stated, a *reliable network flow problem*, one is given a network $N = (V, A, C, K)$ and sets of source nodes and destination nodes, in which each arc in A has a probability associated with it. This is the probability that it would be functional at any point in time. The objective is to compute the probability distribution of network flow characteristics such a maximum flow, or amount of flow transferable within a given budget through the network.[3]

3.5.3 The network cut problem

Consider a road network linking two cities, A and B. The government believes that illegal substances are being smuggled into A from B using this network, and wants to set up a surveillance system to check such smuggling. If the surveillance team is deployed on a road segment, they will check each car passing on that road for illegal substances. Each person in the team can check only one lane of a road at a given time. The government's problem is to find out the minimum number of persons that it needs to put on the team so that all cars moving from B to A are checked at some point on their journey. This problem is called the minimum cut problem.

Formally, in a *minimum cut problem*, one is given a network $N = (V, A, C, K)$, a source node and a destination node. A cut is defined by a subset $S \subseteq A$ such that deletion of S from N creates two components of N, where the source and destination nodes lie in different components. For a given S, let the source node lie in component N_1 and the destination node in component N_2. Let $S_1 \subseteq S$ be a set of arcs with tail in N_1 and head in N_2. The capacity of the cut S is defined as $\sum_{e \in S_1} c_e$. The objective is to find a cut with minimum capacity. Interestingly this problem can be formulated as the dual of the maximum flow problem on the same network.[4] An interesting related problem is the *maximum cut problem* in which the objective is to find a cut with maximum capacity. This problem arises in VLSI design and can be used to

---

[3] For more details on this problem, see M. Rios, V. Marianov, M. Gutierrez, Survivable capacitated network design problem: New formulation and Lagrangean relaxation, Journal of the Operational Research Society 51, (2000), pp.574–582.

[4] For more information on this problem, see P. Elias, A. Feinstein, and C. E. Shannon, Note on maximum flow through a network, IRE Transactions on Information Theory IT-2, (1956), pp.117–119.

model cluster analysis problems. It is more difficult to solve than the minimum cut problem.[5]

3.6 Exercises on Network Flow Problems

Problem 3.1. Transportation Schedules

GTC manufactures its network cables at six different factories in the country. From these factories, the cables are transported to the 44 cable depots, which are warehouses in different parts of the country. From the cable depots the cables are transported to the places were they are actually demanded.

Figure 3.13 contains a schematic road map of 50 locations, labeled 1, ..., 50. Factories where the network cables are manufactured are located in the locations 8, 11, 21, 24, 33, and 36. After production, the cable is stored on spools. The total production of each factory is given in units of ten cable spools. This number is given next to the factory location in Figure 3.13. The remaining 44 locations in Figure 3.13 refer to the 44 cable depots, where the cable spools are stored until they are needed. The demand of the various depots is the number next to the corresponding node in Figure 3.13 (also in units of ten spools, and with a negative sign). The spools are transported by means of trucks which always carry precisely ten spools; the transportation costs per truck, called "truck costs", are shown as numbers next to the corresponding road segments in the network of Figure 3.13.

(a) GTC wants to know a cheapest way of transporting spools to the cable depots in such a way that all depot demands are satisfied.

Since more spools are produced than are demanded, the company also wants to know which factories manufacture spools that are not shipped. How many spools are left at these factories?

(b) There are rumors that, because of the heavy traffic, the government is considering to levy toll on vehicles that use a certain road segment. It is not known yet which segment would be taxed, but road segments $38 \rightarrow 40$, $33 \rightarrow 38$, $12 \rightarrow 10$, $35 \rightarrow 31$, and $29 \rightarrow 34$ are considered as the most likely ones.

GTC is wondering whether the change of one road segment into a toll road will affect the cheapest transportation schedule from Problem 3.3(a). Determine for each of the above five road segments the maximum toll fee that can be charged on that segment without the current transportation plan becoming costlier than some other plan.

(c) There is also a possibility that the direction of traffic on the road segment $33 \rightarrow 34$ will be reversed, i.e., it will become a one-way-traffic from 34 to 33. Determine a cheapest transportation plan under this new circumstance.

---

[5] For more details on this problem, see F. Barahona, M. Grötschel, M. Jünger and G. Reinelt, An application of combinatorial optimization to statistical physics and circuit layout design, Operations Research 3, (1988), pp.493–513.

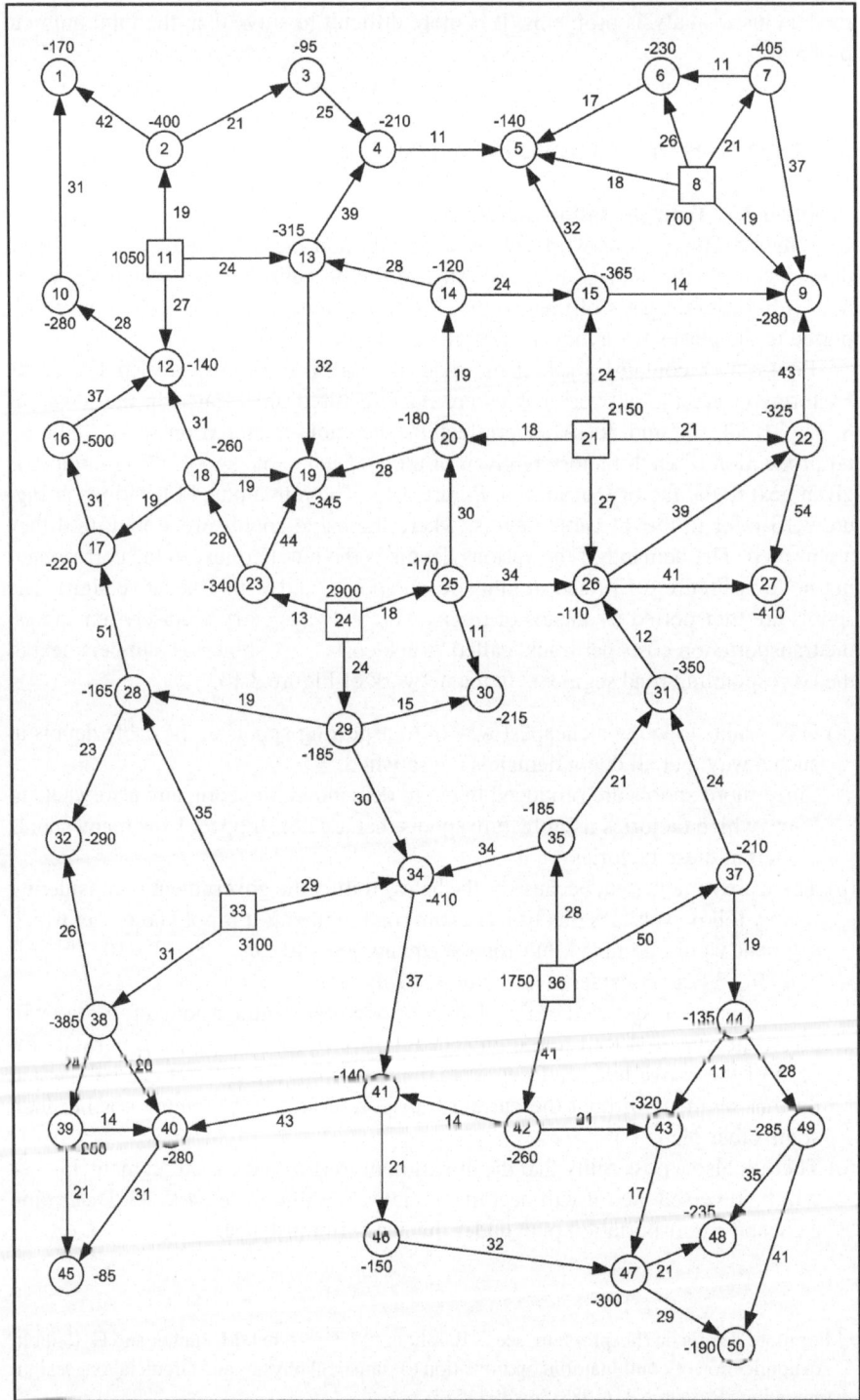

Fig. 3.13. Supplies, demands (in units of 10 spools), and truck costs (in €) on a road map with 50 locations

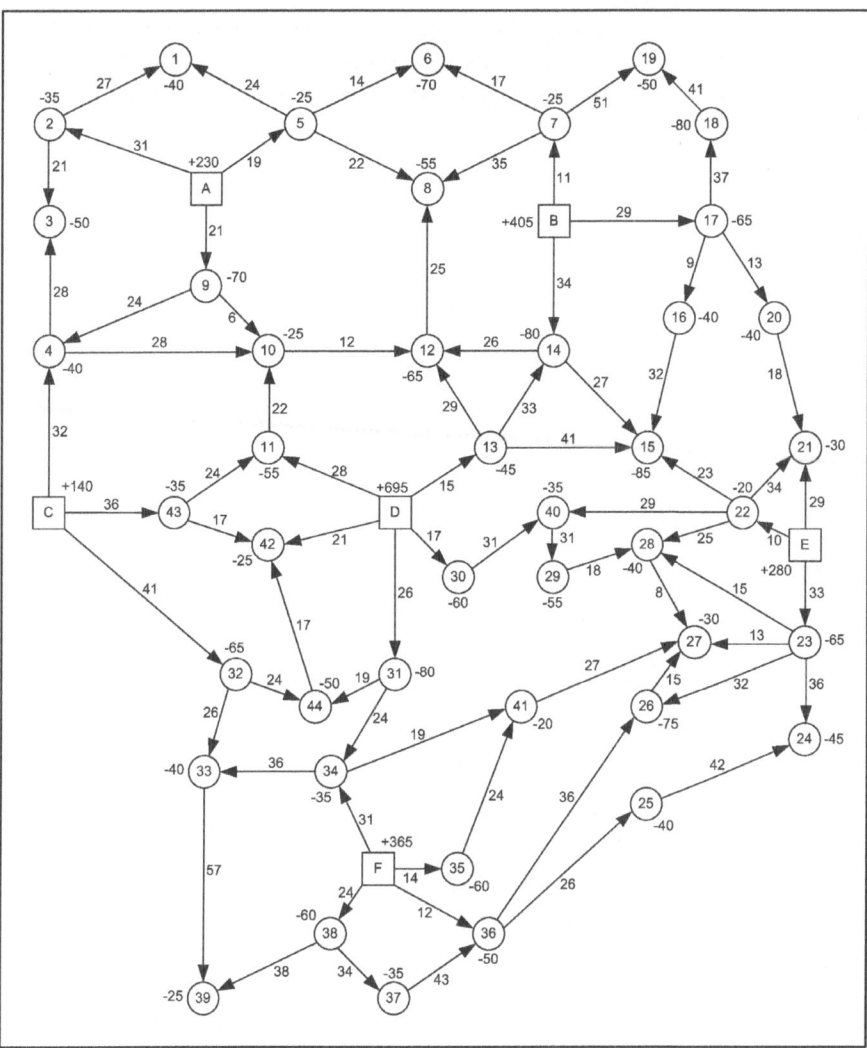

Fig. 3.14. Supplies and demands (in units of 10 spools), and truck costs (in units of €10) on a road map with 50 locations

Problem 3.2. More Transportation Schedules

In another country GTC has to deal with similar questions as in Problem 3.1. In Figure 3.14 the road map for this country is depicted with 50 locations. There are six cable depots, labeled A, B, C, D, E, and F. All numbers in this figure are given in units of ten spools. The inventories at the various depots are the positive numbers next to the depot labels. For instance, in depot C there are 1,400 cable spools in stock. The locations with labels 1, ..., 44 are points where cable is needed, so called demand locations. The numbers next to these labels (with a negative sign) refer to the number of spools demanded at these points. For instance, 650 spools needed at location 17. The numbers attached to the road segments are the transportation costs (in €10 units). For instance, the cost of transporting ten spools with one truck is €290 on the road segment 22→40.

(a) Determine a transportation plan such that all demands are satisfied at minimum truck costs.
(b) It is observed that unacceptable situations occur when trucks arrive from different directions at demand locations. Is it possible to make a feasible transportation plan such that all demands are satisfied and the unacceptable situations are avoided? Explain your answer.
(c) If the answer to the previous question is "no", how will you change the inventories (supplies) in the various depots so that such a plan can be constructed.
(d) Also try to construct a transportation plan without the unacceptable situations, by changing the traffic direction on certain road segments.

Problem 3.3. Vulnerable Connections
A cable network contains two points where two campuses of a university are located. Between these campuses there is a high level of communication that demands a lot of capacity of the cable system. GTC is worried about the connections with an undesirable high utilization degree because these connections are the most vulnerable.

Consider the network of Figure 3.15. This network represents a cable network on 48 locations. Location 1 contains the university campus U1 and location 47 the university campus U2. During the last weeks the communication from U1 to U2 was not without troubles, which were probably caused by an exceptionally high utilization degree of some of the cable connections. GTC wants to know these "bottleneck" connections with very degree of high utilization. The number attached to a cable connection in Figure 3.15 is the estimated flow capacity available for the communication from U1 to U2. The direction of a connection refers to the direction in which there is flow capacity left for the communication from U1 to U2. It is estimated that a total capacity of 900 units is needed in order to satisfy the flow demand from U1 to U2.

(a) Is it possible that a flow of 900 units can be sent through the network from U1 to U2?
    Determine, in case the traffic between the two universities is maximal, the cable connections on which the utilization degree is at least 90%.
    Determine, by inspection, a set of connections, all cables in which have a utilization degree of 100%, for which the total flow on this set is equal to the maximum possible flow from U1 to U2 and that without which, no communication is possible between U1 and U2.
(b) Due to repair activities, the connection unit in location 21 is limited to a flow of only 400 units. Is it still possible to send a flow of 900 units from U1 to U2? Determine the maximum possible flow from U1 to U2 while the repair activities go on. Compare the solution with the one from part (a).

(c) It turns out that in the optimal solution of part (b), the flow on the arc 48→47 is rather high, namely 570 units, so that there is full capacity utilization. The company considers such a bottleneck close to a source location (here 1) or a destination location (here 47) as undesirable, and is thinking about increasing the cable capacity between 48 and 47. Why, in this specific situation, can the cable capacity be increased with an arbitrary amount without changing the maximum possible flow from 1 to 47?

The following problem refers to Problem 1.4 and the remark preceding it; it concerns the R&D of a new type of cable. This cable will be manufactured in a number of factories across the country.

Problem 3.4. Production Planning

One of the cable factories, where the new cable is manufactured (see Problem 1.4), has four identical machines available for the production of the cable. Depending on the fact that different clients demand for different specifications, the company has divided the total demand into 21 different "jobs", each job consisting of a cable demand with a unique specification. For each job, there is a specific release date (the beginning of the day when the job becomes available for processing), and a specific due date (the beginning of the day by which the job must be completed). Each job has a certain processing time (i.e., the number of machine days required for completing the job). Table 3.2 contains the relevant data for the 21 jobs that have to be carried out in the coming planning period of 18 days, namely, for each job the values of the processing time (in machine days), the release date, and the due date are listed. A machine can only work on one job at a time, and each job can be processed by at most one machine at a time. Preemption is possible, meaning that the processing of a job can be interrupted and and can be continued on a different machine on a different day.

(a) Formulate the problem of determining a feasible schedule that completes all jobs after their release dates and before their due dates as a network flow problem.
(b) Determine a solution to the problem from part (a).
(c) At the beginning of this 18 days period, one of the machines breaks down. Is there a feasible schedule if there are three machines during the whole planning period? If not, when must the machine be fixed for a feasible schedule?

Problem 3.5. Department Selection

GTC has obtained an order for modernizing the message transmission system between nine government departments in the state capital. GTC is willing to accept the order only if they can select a subset of the capitals for which the total net profit is maximal. The net profit is the difference between the total cost and the total revenue. The importance of connecting two departments depends on the amount of transmission flow between the departments, and can be seen as a "revenue" of this connection.

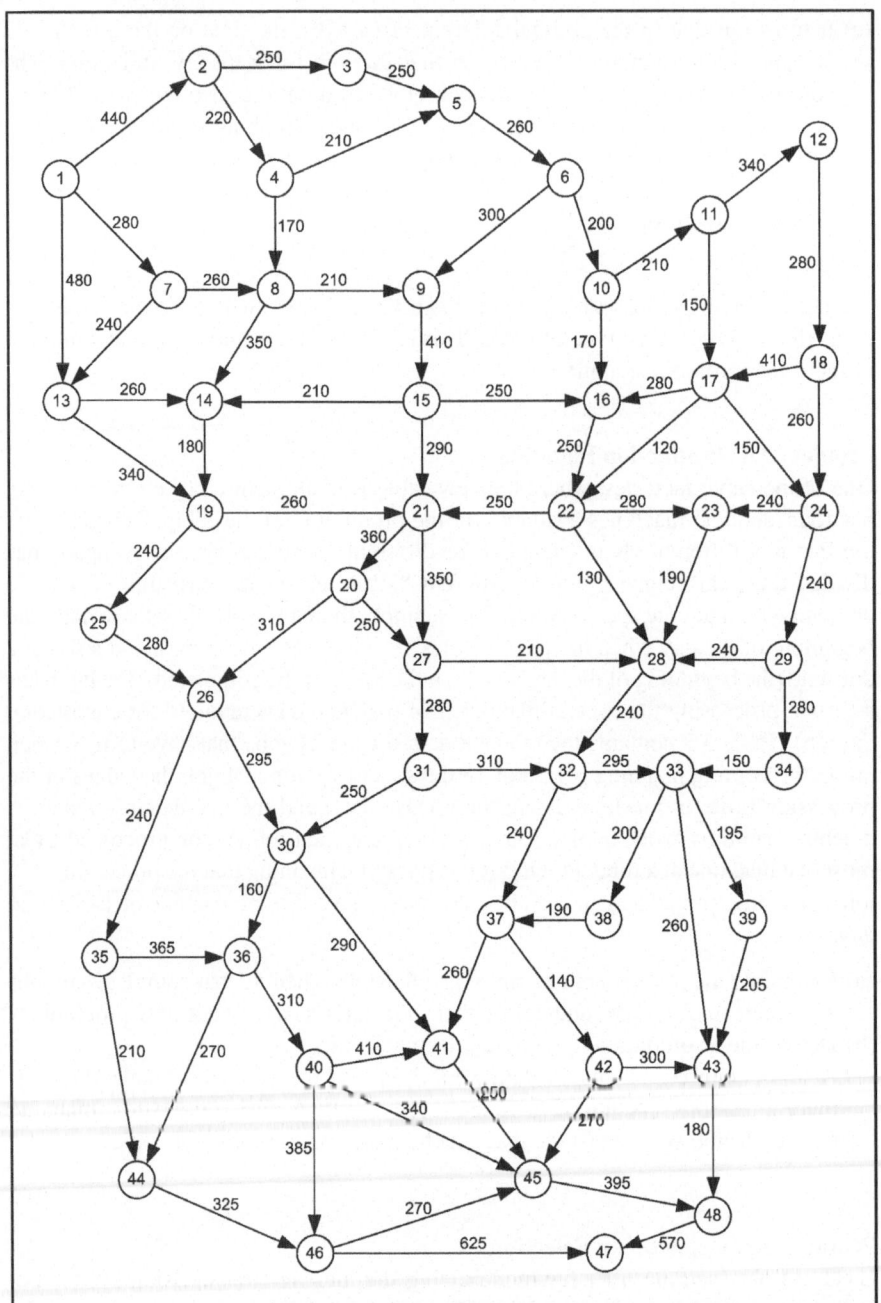

Fig. 3.15. Cable system of Problem 3.3

Table 3.2. Data for Problem 3.4

Job	Processing time	Release date	Due date
1	1.7	3	6
2	2.4	2	6
3	1.2	5	7
4	1.9	1	4
5	2.6	1	5
6	3.0	4	9
7	2.5	3	7
8	3.2	6	11
9	1.8	5	8
10	3.2	7	12
11	2.7	4	7
12	3.0	8	12
13	1.9	7	9
14	2.3	6	9
15	2.8	9	13
16	3.7	9	15
17	2.9	11	15
18	1.6	11	15
19	4.1	12	18
20	3.1	13	18
21	2.4	12	15

For each two departments, the value of the revenue is given in Table 3.3 (in €1,000 units). The cost of including a department into the network is given in the second column of Table 3.3.

Table 3.3. Data for the selection problem of Problem 3.5

Dept.	Including costs	Dept.: 1	2	3	4	5	6	7	8	9
			Revenue (in €1,000 units)							
1	890	—	410	510	70	420	390	180	290	150
2	1,380		—	260	150	270	290	110	440	120
3	780			—	140	450	470	210	350	130
4	1,250				—	140	210	120	90	140
5	1,015					—	380	80	460	110
6	940						—	210	390	260
7	1,190							—	250	280
8	675								—	200
9	980									—

(a) What is the total net profit if all capitals are included in the system?
(b) Which capitals are selected if the total net profit is to be maximized? Compare this answer with the answer to part (a).
(c) Is it possible that the answer to part (b) is not unique, so that there are at least two different sets of capitals with the same optimal net profit? Explain your answer.

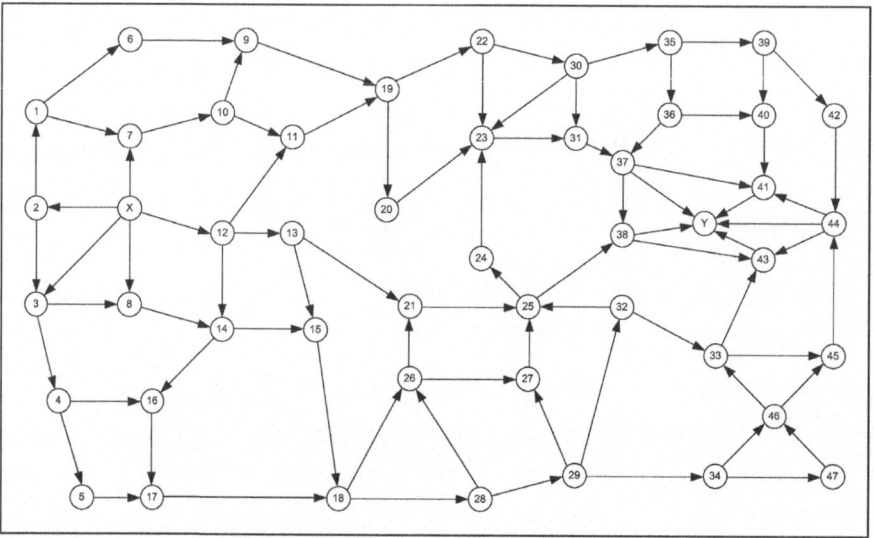

Fig. 3.16. Road Map of the urbanized region of cities X and Y

Problem 3.6. Managing Road Checkpoints
Two large cities X and Y are located on different sides of the state border. The region between X and Y is highly urbanized. The official customs post between the two states is abolished some time ago. However, the governments of both states want to get an idea of the commodity flows from X to Y. To that end they want to open a number of checkpoints along the roads that are used when traveling from X to Y. The road map with the relevant road segments is depicted in Figure 3.16. After careful examination of the commodity flows, it is decided that vehicles traveling only in one direction will be checked, besides the fact that in this heavily urbanized region already many roads are one-way. The road segments and their directions are depicted in Figure 3.16. There is a total of 47 junctions. GTC has obtained the order to build the communication system between the checkpoints. The first question to be solved is the number and the location of the checkpoints. Since the budget for building the communication system is rather restricted and GTC only wants to build a high quality system, the number of checkpoints is rather crucial. Therefore, GTC and the contractor have decided to determine a minimum number of checkpoints that need

to be built. One of the major conditions is that, given the directions of commodity flow, all vehicles that travel from X to Y can be checked. The costs for building each checkpoint and constructing the communication system is estimated at €300,000.

The two governments want to know a minimum price for the construction of a reliable checking system.

(a) Determine the minimum number of checkpoints for the road system of Figure 3.16, together with their locations. What is the minimum amount of money needed for this operation.

(b) Answer the same questions as in part (a), but with the direction of traffic on the road segment 32 − 25 reversed.

(c) Answer the question in part (a), if all flow directions in Figure 3.16 are reversed, and commodity flow from Y to X is considered.

4

Matchings

4.1 Introduction

GTC has five managers Anna, Boris, Caren, Derek, and Elija, labeled A, B, C, D, and E, respectively. It also has five projects labeled P_1 through P_5. Figure 4.1 shows the ability of each of the executives to handle projects — a link between a manager and a project indicates that the manager has the skill set required to handle the project. A

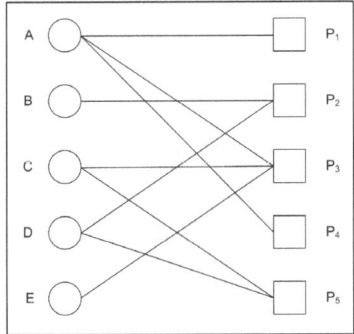

Fig. 4.1. Capabilities of different executives

manager can handle at most one project, and each project, if assigned to a manager, needs to be assigned to exactly one manager. GTC wants to find out how to assign projects to managers such that the maximum number of projects will be assigned.

This problem, and others similar to it in nature in which solutions correspond to pairing of entities are referred to as matching problems. The solutions to these problems are called matchings. Formally stated, *matchings* are subsets of edges in a network such that no two edges in the set are incident on the same node of the network. Thus, a solution to GTC's problem would be the set of edges $\{A - P_3, B - P_2, C - P_5\}$.

G. Sierksma and D. Ghosh, *Networks in Action: Text and Computer Exercises in Network Optimization*, International Series in Operations Research & Management Science 140, DOI 10.1007/978-1-4419-5513-5_7, © Springer Science + Business Media, LLC 2010

However, the matching above is not an optimal solution to GTC's problem. In the problem they require a maximum cardinality matching, also called a *maximal matching*, i.e., a matching which has the maximum cardinality among all matchings. An optimal solution to their problem is the matching $\{A - P_1, B - P_2, C - P_3, D - P_5\}$; see Figure 4.2 in which the thicker lines illustrate the matching. Note that this

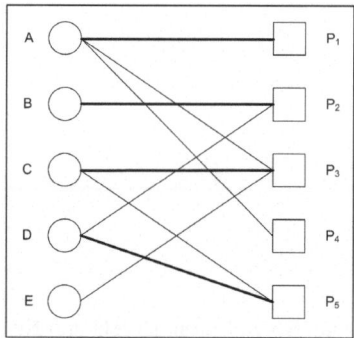

Fig. 4.2. A maximal matching

is one of several maximal matchings possible. Also note that even though there are five managers and five projects, the best solution here is able to allocate managers to four projects only. This is because Anna is the only person who is competent to be assigned to either P_1 or P_4, and the project that she is not assigned to remains unassigned in the solution. A matching, such that for each node in the network exactly one edge in the matching is incident on it, is called a *perfect matching*.

Next, consider approaches in which choosing one matching over another is not determined by the cardinality of the matching alone. To make such choices, one needs to attach weights on the edges in the network. For example, in addition to having the required skill set for being assigned to a project, suppose that there is a measure of the suitability of a manager for a project (where a higher value indicating a higher suitability), and one wants to find the best way of matching managers to projects. The problem is depicted by a weighted graph as in Figure 4.3. The weights on the edges indicate the measure of suitability of managers to projects. Combinations that are not represented in the network are assumed to have extremely low measures. In this case, the objective is to obtain a *maximum weight matching*, i.e., a matching in which the sum of the edge weights is maximum. In situations where the underlying graph is bipartite, such a maximum weight matching problem is called an *assignment problem*. In this example, the matching $\{A - P_4, C - P_5, D - P_3, E - P_3\}$ is the best assignment possible.

In the problems described so far in this chapter, the underlying networks are bipartite. It does not make sense in the problems discussed earlier to assign a project to another project, or a manager to another manager. However, not all matching problems are defined on bipartite graphs.

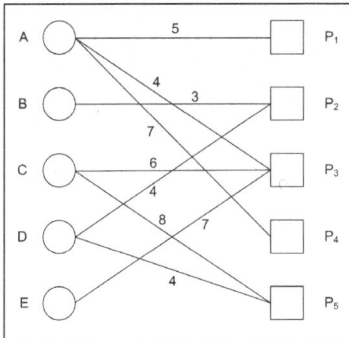

Fig. 4.3. A weighted matching problem

Consider for example, that seven managers (labeled A, B, ..., G) need to form two member teams to deal with projects in the departments. Teams can work only when the two members in the team get along with each other. The inter-personal relations between managers are depicted in the left-hand side of Figure 4.4 by a network in which managers are represented by nodes, and there is an edge between two nodes if the managers they represent get along well with each other. For instance, manager A can get along with one among B, D, F, and G, but not with C or E. This network is obviously not bipartite. However, we have a matching problem since a solution would be a subset of the edges, no two of which are incident on the same node. This problem is known in the literature as the *roommate problem*. A solution

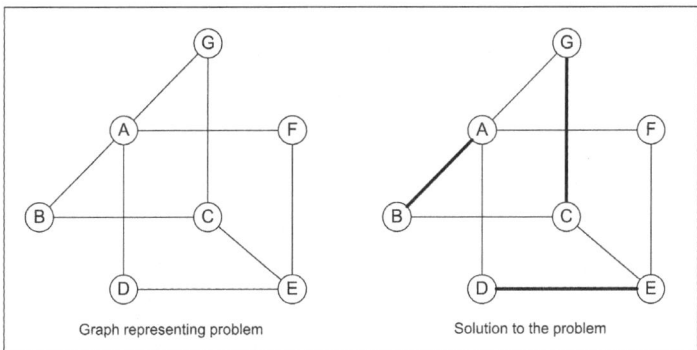

Fig. 4.4. A matching problem on a non-bipartite graph and its solution

to the problem (shown in the right-hand side of Figure 4.4) is the subset of edges {A – B, C – G, D – E}. F is unable to team up with anyone in this solution. Note that this is an optimal solution, but not the only optimal solution for this problem. Another optimal solution for example would be the set {A – G, B – C, E – F}, which leaves D without a teammate.

The last type of matching problems described here is called the *bottleneck matching problem*. Consider the assignment problem discussed earlier and illustrated using Figure 4.3. Assume that the objective is to find a maximum cardinality assignment of managers to projects, but to ensure such that the minimum value of the capability of an manager assigned to a particular project is the largest possible. This ensures that the "worst" of the assignment decisions is the best possible. This "worst" assignment decision (the least cost edge) is referred to as the *bottleneck*. In the optimal assignment in the example for instance, the bottleneck is edge $D - P_2$, with a value of 4. By inspection, it is easy to see that this assignment is also optimal for the bottleneck matching problem.

4.2 Applications

Matching problems have varied applications in addition to the ones stated in the introductory session. Some of the other applications are sketched in this section.

4.2.1 Constructing university timetables

Consider a term at an university, in which classes have to be scheduled for students. Each student opts for multiple courses and the problem is to assign course class slots to classrooms. For this, from the student options, classes are constructed. Each class has a set of students and an instructor. Classroom slots have a physical classroom and a time-interval. A bipartite network is constructed from this data in which classes make up one partition and classroom slots make up the other partition. Classes and classroom slots are connected to ensure that if a class is connected to a slot, then the classroom in the slot is large enough to accommodate the class, and no other class connected to the slot has the same instructor or has students who are also in this class. The classroom slot allotment problem then reduces to a maximum cardinality matching problem in this network, to ensure that the maximum number of classes can be accommodated to form a timetable.

4.2.2 Constructing 3-dimensional models from 2-dimensional data

In image processing, an interesting problem is to construct 3-dimensional models from information provided in two dimensions. For example, given two images of the same solid object in two dimensions taken from slightly different viewpoints, one needs to compute its model in three dimensions. The way in which this is done is to find out the position of the same feature of the object (for example a corner of the object) in the two images, and given the positions of the viewpoints from which the two images were taken, find the position of the feature in 3-dimensional space using triangulation. However, to follow this method, first one has to find out which feature in one image corresponds to which feature in the second image. This problem is a weighted bipartite matching problem. The exact method of constructing a bipartite network from such images is beyond the scope of the book[1].

4.2.3 Crew pairing

Consider the problem of the management in an airline which has to form pilot co-pilot pairings among candidates who speak different languages. The airline wants to construct teams where the two members speak the same language. In addition to ensuring that the pilot and the co-pilot speak the same language, the airline wants to create pairings in which the pilot and the co-pilot get along well with each other. This problem can be solved using a weighted non-bipartite matching problem. A network is constructed in which the nodes correspond to candidates, and two candidates are connected to each other if they speak the same language. The weight of an edge is a measure of the interpersonal relations between the pair of candidates that it joins, a high value indicating a high level of compatibility. The solution to the problem is then to construct a maximum weight matching in this network and to form pairings of candidates who share the same edge in the optimal solution.

4.3 Linear Programming Formulations

All the problems described in the previous section can be solved through the linear programming technique. For each edge $i - j$ in the network, in each of the formulations, define a decision variable x_{ij} that assumes a value of 1 if the edge is part of an optimal matching, and 0 otherwise. A common set of constraints that define matchings in all the formulations ensure that each node in the network has at most one edge in the matching incident on it. So we define the constraint set

$$\sum_{j:i-j\in E} x_{ij} \leq 1 \text{ for each } i \in V. \qquad (4.1)$$

The special structure of this set of constraints ensures that a relaxation to the constraint set, in which the binary decision variables are relaxed to include all real values between 0 and 1, has integer corner points. Therefore, for matching problems that have only constraints of the form (4.1) in their can be solved as linear programs with the x_{ij} variables restricted to values between 0 and 1, instead of binary linear programs[2]. Additionally, one does not need to add the constraints restricting the decision variables not to exceed 1, since the constraints of type (4.1) along with non-negativity constraints are sufficient to ensure this condition.

In the remaining part of this section linear programming formulations will be developed for the four types of matching problems introduced in the previous section.

---

[1] For more details on this problem, see D.B. Goldgof, H. Lee, and T.S. Huang, Matching and motion estimation of three-dimensional point and line sets using eigenstructure without correspondences, Pattern Recognition, 25, (1992), 271–286.

[2] See G. Sierksma, Linear and Integer Programming: Theory and Practice, Chapter 7.

4.3.1 The maximum cardinality matching problem

In the maximum cardinality matching problem, one is interested in finding a matching with the largest possible number of edges in the matching. So the objective function is simply the sum of the number of edges in the matching, and the objective is mathematically represented as

$$\text{Maximize} \sum_{i-j \in E} x_{ij}.$$

The only set of constraints required are of type (4.1).

A complete formulation of a linear program to construct a maximum cardinality matching in a network $N = (V, E)$ is thus the one shown in Figure 4.5.

Maximize

$$z = \sum_{i-j \in E} x_{ij}$$

Subject to

$$\sum_{j:i-j \in E} x_{ij} \leq 1 \quad \text{for each } i \in V$$

$$x_{ij} \geq 0 \quad \text{for each } i - j \in E$$

Fig. 4.5. Linear programming formulation of the maximum cardinality matching problem

Figure 4.6 shows the formulation of the maximum cardinality matching problem for the network in Figure 4.1. For notational convenience, the nodes for projects P_1 through P_5 are labeled 1 through 5.

4.3.2 The maximum weight matching problem

In the maximum weight matching problem, one is interested in finding a matching for which the sum of weights of edges in the matching is the maximum possible. The objective function in a linear programming formulation for this problem is therefore a weighted sum of edges in the solution, and the objective is mathematically represented as

$$\text{Maximize} \sum_{i-j \in E} w_{ij} x_{ij},$$

where w_{ij} is the weight assigned to the edge $i - j$ in the network. As in the maximum cardinality matching problem, the only set of constraints required are of type (4.1).

A complete formulation of a linear program to construct a maximum weight matching in a network $N = (V, E)$ is thus the one shown in Figure 4.7.

Maximize

$$z = x_{A1} + x_{A3} + x_{A4} + x_{B2} + x_{C3} + x_{C5} + x_{D2} + x_{D5} + x_{E3}$$

Subject to

$$x_{A1} + x_{A3} + x_{A4} \leq 1 \quad \text{(Constraint (4.1) at A)}$$
$$x_{B2} \leq 1 \quad \text{(Constraint (4.1) at B)}$$

There are three more constraints for nodes C, D, and E.

$$x_{A1} \leq 1 \quad \text{(Constraint (4.1) at } P_1)$$
$$x_{B2} + x_{D2} \leq 1 \quad \text{(Constraint (4.1) at } P_2)$$

There are three more constraints for nodes P_3, P_4, and P_5.

$$x_{A1}, x_{A3}, x_{A4}, x_{B2}, x_{C3}, x_{C5}, x_{D2}, x_{D5}, x_{E3} \geq 0 \quad \text{(Nonnegative flows)}$$

Fig. 4.6. Formulation of the maximum cardinality matching problem for the network in Figure 4.1

Maximize

$$z = \sum_{i-j \in E} w_{ij} x_{ij}$$

Subject to

$$\sum_{j:i-j \in E} x_{ij} \leq 1 \quad \text{for each } i \in V$$
$$x_{ij} \geq 0 \quad \text{for each } i - j \in E$$

Fig. 4.7. Linear programming formulation of the maximum weight matching problem

Figure 4.8 shows the formulation of the maximum weight matching problem for the network in Figure 4.3. Here too, for notational convenience, the nodes for projects P_1 through P_5 are labeled 1 through 5.

4.3.3 The non-bipartite matching problem

The linear programming formulations described in the previous two cases do not assume that the network being analyzed is bipartite. Therefore the formulation de-

Maximize

$$z = 5x_{A1} + 4x_{A3} + 7x_{A4} + 3x_{B2} + 6x_{C3} + 8x_{C5} + 4x_{D2} + 4x_{D5} + 7x_{E3}$$

Subject to

$$x_{A1} + x_{A3} + x_{A4} \leq 1 \quad \text{(Constraint (4.1) at A)}$$
$$x_{B2} \leq 1 \quad \text{(Constraint (4.1) at B)}$$

There are three more constraints for nodes C, D, and E.

$$x_{A1} \leq 1 \quad \text{(Constraint (4.1) at } P_1)$$
$$x_{B2} + x_{D2} \leq 1 \quad \text{(Constraint (4.1) at } P_2)$$

There are three more constraints for nodes P_3, P_4, and P_5.

$$x_{A1}, x_{A3}, x_{A4}, x_{B2}, x_{C3}, x_{C5}, x_{D2}, x_{D5}, x_{E3} \geq 0 \quad \text{(Nonnegative flows)}$$

Fig. 4.8. Formulation of the maximum weight matching problem for the network in Figure 4.3

veloped to solve the maximum cardinality matching problem can immediately be applied to solve the roommate problem. As an example, a linear programming formulation for the roommate problem (depicted in Figure 4.4) is shown in Figure 4.9.

Maximize

$$z = x_{AB} + x_{AD} + x_{AF} + x_{AG} + x_{BC} + x_{CE} + x_{CG} + x_{DE} + x_{EF}$$

Subject to

$$x_{AB} + x_{AD} + x_{AF} + x_{AG} \leq 1 \quad \text{(Constraint (4.1) at A)}$$
$$x_{AB} + x_{BC} \leq 1 \quad \text{(Constraint (4.1) at B)}$$

There are five more constraints for nodes C, D, E, F, and G.

$$x_{AB}, x_{AD}, x_{AF}, x_{AG}, x_{BC},$$
$$x_{CE}, x_{CG}, x_{DE}, x_{EF} \geq 0 \quad \text{(Nonnegative flows)}$$

Fig. 4.9. Formulation of the roommate problem for the network in Figure 4.4

It should be immediately obvious that a weighted version of the roommate problem can be formulated exactly in the same way as the weighted matching problem.

4.3.4 The bottleneck matching problem

Bottleneck matching problems are more difficult to formulate than the other matching problems. The solution to such a problem is a maximal matching. So one needs to first find the cardinality of a maximal matching for the problem. Once the cardinality of the maximal matching is obtained, one then has to evaluate all maximal matchings and find the one for which the weight of the bottleneck edge is as high as possible.

An easy way of computing the size of the maximal matching in a network $N = (V, E, w)$ is to solve a maximum cardinality problem on the network using, for example, the linear programming formulation for maximum cardinality matchings as described earlier in this section, and measuring the cardinality of the solution output. Let us assume that in the network, the maximum cardinality matching has cardinality m.

Next, the problem of finding the best matching among the matchings with cardinality m $(m \geq 1)$ has to be formulated. In order to achieve this, constraints are added to ensure that the solution output by the formulation is a matching. This is easily done by including constraints of the form (4.1) for all nodes in the network. Next, a constraint is added to restrict the solution space to the set of maximum cardinality matchings. This is achieved with the constraint

$$\sum_{i-j\in E} x_{ij} = m. \tag{4.2}$$

Finally, in order to minimize the weight of the bottleneck edge one defines a variable v to store the weight of the bottleneck edge. Constraints are added to ensure that v does not take a value that is larger than the weights of any of the edges in the solution output. This is done by modeling the statement

if $x_{ij} = 1$, then $w_{ij} \geq v$ for each $i - j \in E$.

This condition is implemented using the constraint

$$w_{ij} - v + M(1 - x_{ij}) \geq 0 \text{ for each } i - j \in E, \tag{4.3}$$

where M is a number larger than $\max_{i-j\in E} w_{ij} + 1$. If $x_{ij} = 1$ for any edge $i - j$ in the network, then for that edge this constraint evaluates to $w_{ij} \geq v$. On the other hand, if $x_{ij} = 0$, then this constraint evaluates to $w_{ij} \geq v - M$, a redundant constraint. Unfortunately, the addition of the constraints (4.2) and (4.3) in the constraint set destroys the property that enabled the replacement of the binary valued x_{ij}'s with continuous variables. Hence, in the formulation of bottleneck matching problems the x_{ij}'s must be considered to be binary variables rather than continuous ones.

The objective of the bottleneck matching problem is simply

Maximize v.

A complete formulation of a bottleneck matching problem on a network $N = (V, E)$ is given in Figure 4.10. In the formulation it is assumed that the cardinality m of the maximum cardinality matching in N is known.

Maximize

$$z = v$$

Subject to

$$\sum_{i-j \in E} x_{ij} = m$$

$$w_{ij} - v + M(1 - x_{ij}) \geq 0 \qquad \text{for each } i - j \in E$$

$$v \geq 0$$

$$x_{ij} \in \{0, 1\} \quad \text{for each } i - j \in E$$

Fig. 4.10. Formulation of the bottleneck matching problem

Figure 4.11 shows the formulation of the bottleneck matching problem on the network shown in Figure 4.3. The value of m is taken to be 4 since the maximum cardinality matching on this network has four edges. We may choose $M = 9$, since the maximum weight of an edge in the network is 8.

4.4 Algorithms for Matchings

There are several algorithms that solve large scale matching problems more efficiently than linear programming. However, most of these algorithms are complex to describe, and their description is beyond the scope of this book. In this section we describe one algorithm which is easy to understand, namely an algorithm for finding the maximum cardinality matching problem on bipartite graphs.

The maximum cardinality matching problem on a bipartite graph can be shown to be identical to the maximum flow problem in an auxiliary network. So we can solve it efficiently in three steps. For this illustration, let us consider a bipartite graph (V, E) in which the node set V is partitioned into disjoint sets V_1 and V_2.

In the first step an auxiliary network is defined. The node set of this network has all nodes of the original network, in addition to a dummy source node and a dummy destination node. The set of arcs in the auxiliary network has three sets of arcs. In the first set are unit capacity arcs that go from the source node to all nodes in the first subset of the partition. The second set of arcs contains unit capacity arcs from all nodes in the second subset to the destination node. The third set of arcs contains arcs corresponding to edges in the original graph. For every edge $i - j$ in the original network in which $i \in V_1$ and $j \in V_2$, we include a unit capacity arc $i \rightarrow j$ in the auxiliary network.

In the second step, the maximum flow from the source node to the destination node in the auxiliary network is obtained. This is done using one of the more efficient algorithms like the Ford-Fulkerson algorithm described in Chapter 3. This results in

Maximize

$$z = v$$

Subject to

$$x_{A1} + x_{A3} + x_{A4} + x_{B2} + x_{C3} + x_{C5} + x_{D2} + x_{D5} + x_{E3} = 4 \quad \text{(Condition (4.2))}$$

$$5 - v + 9(1 - x_{A1}) \geq 0 \quad \text{(Constraint (4.3) for A – 1)}$$
$$4 - v + 9(1 - x_{A3}) \geq 0 \quad \text{(Constraint (4.3) for A – 3)}$$

There are seven more constraints for the other edges in the network.

$$x_{A1} + x_{A3} + x_{A4} \leq 1 \quad \text{(Constraint (4.1) at A)}$$
$$x_{B2} \leq 1 \quad \text{(Constraint (4.1) at B)}$$

There are three more constraints for nodes C, D, and E.

$$x_{A1} \leq 1 \quad \text{(Constraint (4.1) at 1)}$$
$$x_{B2} + x_{D2} \leq 1 \quad \text{(Constraint (4.1) at 2)}$$

There are three more constraints for nodes 3, 4, and 5.

$$v \geq 0 \quad \text{(Continuous variable)}$$
$$x_{A1}, x_{A3}, x_{A4}, x_{B2}, x_{C3}, x_{C5}, x_{D2}, x_{D5}, x_{E3} \in \{0, 1\} \quad \text{(Binary decision variables)}$$

Fig. 4.11. Formulation of the bottleneck matching problem on the network in Figure 4.3

an optimal solution to the maximum flow problem in which all arcs that carry flows carry one unit of flow.

In the third and final step, the optimal solution obtained in the second step is used to obtain an optimal solution to the maximum cardinality matching problem. This involves concentrating on the arcs in the third set defined in the first step. If and only if the arc $i \rightarrow j$ in the auxiliary network carries flow in the optimal solution obtained in the second step, then the corresponding edge $i - j$ in the original network is part of the optimal matching. Since each arc from the source node to a node in the first set has unit capacity, at most one of the arcs from each node in the first set can carry unit flow in the solution obtained in the second step. Similarly, since the capacity of each arc from the nodes in the second set to the destination node has unit capacity, only one of the arcs to each node in the second set can carry unit flow. Thus, the flow pattern represented in the optimal solution to the maximum flow problem described in the second step corresponds to a matching. Since the objective is to maximize the flow from the source node to the destination node, an optimal flow would have the maximum possible number of arcs between the two sets that carry unit flows.

As an example, consider the matching problem described by the network in Figure 4.1. The aim is to obtain a maximum cardinality matching in this network. The auxiliary network for the network in Figure 4.1 is constructed by adding a dummy source node S and a dummy destination node T to the set of nodes and creating the three sets of arcs as described earlier in this section. The auxiliary network thus formed is shown in Figure 4.12. Arc capacities and costs have not been printed along the arcs, but each of the arcs have unit capacity, and arc costs are not relevant to the solution process.

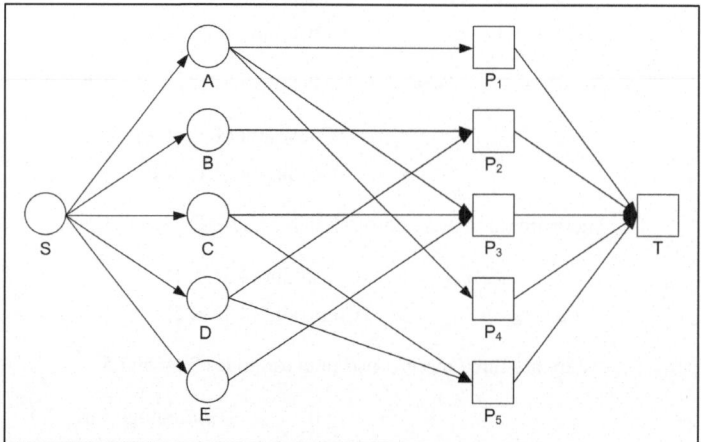

Fig. 4.12. Converting a matching problem to a flow problem

In the second step the maximum flow problem from the source node S to the destination node T through the network of Figure 4.12 is solved. An optimal solution to this problem is shown in Figure 4.13, in which the thicker lines correspond to the arcs that carry unit flow.

In the third step, an optimal solution to the network shown in Figure 4.1 is obtained. In the solution to the problem in the second step, arcs $A \rightarrow P_1$, $B \rightarrow P_2$, $C \rightarrow P_3$, and $D \rightarrow P_5$ carry unit flows. Therefore a maximum cardinality matching in the network shown in Figure 4.1 is the set $\{A \rightarrow P_1, B \rightarrow P_2, C \rightarrow P_3, D \rightarrow P_5\}$.

4.5 Other Matching Problems

4.5.1 The stable marriage problem

A medical school has ten interns and wants to allocate them to ten hospitals. Each intern has a priority ordering of the hospitals, and would prefer to join a hospital higher in her ordering if she is allowed to. The hospitals also have a ranking of interns, and a particular hospital would be willing to trade an intern with another if

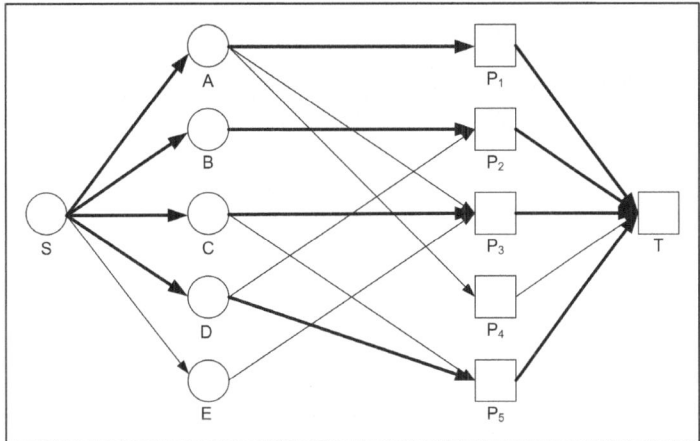

Fig. 4.13. Solution to the maximum flow problem of Figure 4.12

the former intern is ranked higher in its ranking. Consider a situation where an intern A is assigned to a hospital P, and another intern is assigned to a hospital Q. If A prefers Q to P, B prefers P to Q, P prefers B to A, and Q prefers A to B, then the assignment is said to be unstable, since it would be better all round if A was assigned to Q, and B to P. A matching of interns to hospitals is called stable if there are no unstable assignments of the type described above. The problem of finding a stable assignment of interns to hospitals is called the *stable marriage problem*.[3]

There are other problems related to the stable marriage problem. One of them is called the *college admissions problem*. This problem differs from the stable marriage problem in that each college can accept more than one applicant, while the stable marriage problem allows each hospital to accept exactly one intern.

4.5.2 The 3-dimensional stable matching problem

Consider a school that has a policy of admitting equal numbers of domestic students and foreign students. In order to ensure that these students understand the nuances of multiple cultures, the school also requires each domestic student pairs up with a foreign student in rented apartments. Each domestic student has a ranking of each of the foreign students that she wants to pair up with, and so does each of the foreign students. Hence the preferences are mentioned in the form "Domestic student x would like to pair up with foreign student y in apartment z.". The problem is to find a stable matching of domestic and foreign students, and apartments. This problem is called a *3-dimensional stable matching problem*[4].

---

[3] For more details on this problem, see D. Gusfield and R.W. Irving, The Stable Marriage Problem: Structure and Algorithms, MIT Press, 1989.

[4] There is literature to show that not all 3-dimensional stable matching problems have solutions (see for example, A. Alkan, Non-existence of stable threesome matchings, Math-

4.6 Exercises on Matching Problems

Problem 4.1. Scheduling Technicians

GTC employs fifteen maintenance technicians who are used to solve problems that customers have with their communication devices. The technicians are paid by hours of work. All technicians have the same wage per hour, but are not equally qualified on all problems. The problems are divided into ten different categories. This morning, twenty customers have problems in the categories shown in Table 4.1. However only fifteen customers can be served, because each technician is assigned to only one job. The twenty customers have called in the order 1 through 20, meaning that Customer 1 has called first, and Customer 20 called last. The time required by a tech-

Table 4.1. Customers and their categories

Customer	Category	Customer	Category	Customer	Category
1	1	8	2	15	7
2	6	9	7	16	9
3	3	10	4	17	4
4	7	11	5	18	7
5	5	12	3	19	2
6	8	13	9	20	1
7	9	14	10		

nician to solve a problem is called the repairing time. Assume that GTC knows the repairing times per category and per technician. These times are shown in Table 4.2. For example, Technician 2 needs 17 minutes to solve a problem in Category 1. GTC wants to serve fifteen customers at minimum technician wage cost.

(a) Assume that the customers are served on a first-call-first-served basis.
  1. By inspection, how would you assign the fifteen technicians to the customers? What is the rationale behind your solution procedure?
  2. Explain why your solution need not be optimal.
  3. What is an optimal assignment of technicians to customers, and what is the total repairing time?
(b) GTC now wants to analyze the consequences of serving fifteen customers out of twenty for which the total wage is as low as possible.
  1. By inspection, how would you assign the fifteen technicians to the customers? What is the rationale behind your solution procedure?
  2. Explain why your solution need not be optimal.
  3. Set up an optimization model to solve this problem.

---

ematical Social Sciences 16, (1988), pp.207–209), but under certain preference orderings, they always do have solutions (see E. Boros, V. Gurvich, S. Jaslar, and D. Krasner, Stable matchings in three-sided systems with cyclic preferences. Discrete Mathematics 286, (2004), pp.1–10).

Table 4.2. Repairing times per technician per category

Technicians	Category									
	1	2	3	4	5	6	7	8	9	10
1	26	76	159	187	41	45	193	49	174	201
2	17	91	128	162	50	42	167	38	146	122
3	28	111	152	145	51	52	212	44	156	221
4	24	84	155	209	58	44	146	41	122	177
5	26	85	139	201	66	51	144	44	164	190
6	21	95	176	197	65	58	196	52	157	209
7	19	96	173	183	54	50	192	50	97	192
8	19	84	135	177	52	45	175	42	99	212
9	17	98	119	178	58	46	194	38	178	158
10	24	80	109	199	60	50	218	61	142	150
11	18	88	146	178	57	48	159	48	156	190
12	28	105	124	158	55	43	150	36	132	183
13	22	102	129	157	48	57	233	50	158	155
14	28	110	168	217	48	58	159	47	136	179
15	24	98	123	181	54	49	112	44	196	135

    4. What is an optimal assignment of technicians to customers now, and what is the total repairing time?

(c) Why is the optimal solution to part (b) not worse than the optimal solution to part (a) in terms of repair times?

(d) Draw and analyze the perturbation function corresponding to the model used in part (b), of the coefficient representing the time that Technician 2 needs to serve Customer 14[5].

(e) Based on the first-call-first-serve principle, what happens to the optimal solution to part (a) when Technician 2 is not available and Technician 8 visits two customers?

(f) What happens to the optimal assignment of part (b) if Customer 4 solves the problem himself, so that no technician is needed for this customer?

Problem 4.2. Executing Two-Stage Projects
GTC has ten specialized departments all over the world. There are 35 projects to be executed during the next six months. Each project consists of two consecutive stages. The first stage of each project takes three months. After these three months, the next stage of the projects starts, taking again three months. Departments are assigned to project stages. The assignments are listed in Table 4.3. For example, in the first stage, department 1 (D1) is assigned to the projects with labels 1, 3, 6, 10, 19, 26, and 34. In the second stage, D1 is assigned to the projects with labels 6, 9, 12, 16, 19, and 23.

---

[5] The *perturbation function* of a model parameter is a function of which the values are the objective values of the model for all feasible values of that parameter.

Table 4.3. Assignment of departments to project stages

Department	Stage	Projects
D1	S1	1 3 6 10 19 26 34
	S2	6 9 12 16 19 23
D2	S1	12 16 18 19 20 21 27
	S2	3 5 11
D3	S1	14 15 22 24 29
	S2	1 7 8 16 19 23 24 29 34
D4	S1	3 5 6 7 11 13 17
	S2	2 4 8 13 17 21 22
D5	S1	2 4 8 9 23 25 30 33 35
	S2	10 14 18 20 25 27 28
D6	S1	1 9 13 16 21 26 31 34
	S2	12 15 18 26 30
D7	S1	1 6 9 26 31
	S2	19 26 31 35
D8	S1	3 15 16 21 22 23 29 33 34 35
	S2	5 9 10 16 17 27 33
D9	S1	5 6 9 13 14 19 23 26
	S2	1 6 8 15 23 29 30 32
D10	S1	9 11 18 21 25 29 34
	S2	6 10 16

After the first stage, a department can stay with the same project. For example, D1 participates in Project 6 during all six months.

Starting a new second stage project needs preparation time. In order to limit the preparation time, each department opens a help desk at the beginning of the second stage. In the help desk of, say D1, D1 may help departments that are involved in second stage projects that D1 was involved with during the first stage. However, due to the complexity of the projects, a help desk can only help one department, while a department is allowed to use only one help desk. Note that a second stage department can be helped on more than one project. For example, the help desk of D1 (denoted by H1) can inform D3 on the three projects, 1, 19, and 34. It is assumed that all preparation times are equal, and all help desks require equal times per project. GTC wants to minimize the total preparation time.

For your convenience, Table 4.4 partly lists the number of new projects on which a help desk can help a department.

(a) Calculate the missing figures in Table 4.4.
(b) By inspection, how would you assign the help desks to the departments? What is the rationale behind your solution procedure? What is the total number of projects on which help desks help departments in your solution? List all projects that are served by the help desk.

Table 4.4. Number of new projects for departments

	Department
Helpdesk	1 2 3 4 5 6 7 8 9 10
1	0 1 3 0 1 0 1 1 ? 2
2	2 0 2 1 3 2 1 2 0 1
3	0 0 0 1 1 1 0 0 2 0
4	0 ? 1 0 0 0 0 2 0 1
5	2 0 2 1 0 1 1 1 2 0
6	2 0 ? 2 0 0 0 1 1 1
7	1 0 0 0 0 0 0 1 1 1
8	2 1 3 2 0 ? 1 0 2 1
9	2 1 2 1 1 1 1 ? 0 1
10	1 1 1 1 1 1 0 1 1 0

(c) Find an assignment that results in the minimum preparation time, and list all projects that are served by the help desks. Compare this answer to the answer of part (b).

During the second stage, the four projects labeled 11, 12, 19, and 21 need new software. The company that implements the software needs information about each of these projects, because each project needs a different kind of software. GTC has decided that the software company can use only one help desk, and that a help desk may serve either a department or the software company.

(d) If the software company chooses the help desk of which the number of projects served was maximal, what would then be the optimal solution? List all projects that are served by the help desks in your solution.

(e) What is an optimal assignment of help desks to departments and the software company? List all projects that are served by the help desks. Compare this answer to the answer of part (d).

(f) Projects 1 and 16 turn out to be very costly. GTC decides to stop these projects after the first stage. Give an optimal assignment of the help desks to the departments after this decision. Again, list all projects that are served by the help desks. Compare this answer to the answer to part (e).

Problem 4.3. Team Building Excursion

Once a year the department located in Brussels organizes an excursion for its 40 employees. This excursion is organized in a holiday resort in the forest. The employees leave on Friday and come back on the next Sunday. The goal of the excursion is team building. Therefore, not everything in the holiday resort is included in the service provided by the resort. All the tasks to be done by the employees are listed in Table 4.5. For example, during the excursion there are two days with breakfast. Each of these days the breakfast is divided into two subtasks. So, a total of four tasks is needed to prepare breakfast.

Table 4.5. Tasks to be done by employees

Task Description	# Subtasks	# Days
1. Preparing breakfast	2	2
2. Setting the tables (breakfast)	3	2
3. Doing the dishes (breakfast)	3	2
4. Preparing lunch	2	3
5. Preparing supper	5	3
6. Setting the tables (supper)	4	3
7. Doing the dishes (supper)	6	3
8. Cleaning the rooms on the last day	13	1

An overall total of 80 subtasks is needed to fully cover all tasks. The excursion committee has decided that all 40 employees should do exactly two subtasks.

To ease the pain of doing "forced labor", employees may label the tasks as "doable" or "non-doable". That is, if an employee labels a task as doable, then he/she is willing to do all subtasks of the task. The doable tasks of each employee are shown in Table 4.6. For example, Employee 1 is willing to do tasks 3 and 5. The excursion

Table 4.6. Doable tasks for each employee

Employee	Doable Tasks	Employee	Doable Tasks
1	3, 5	21	5, 6
2	1 ,2	22	none
3	6	23	6
4	2	24	1, 3
5	7	25	3, 5
6	3	26	4
7	3, 4	27	5, 6
8	2, 3	28	6
9	7	29	6, 7
10	none	30	6, 7
11	6, 7, 8	31	6, 7
12	2	32	none
13	8	33	4, 8
14	2	34	none
15	6, 7	35	3
16	4	36	5
17	4	37	5
18	6, 7	38	7
19	4, 8	39	none
20	3	40	4

committee wants to assign employees to tasks such that the number of non-doable tasks is minimized. It is assumed that the workload of all subtasks is equal.

(a) By inspection, how would you assign the employees to the tasks? How many non-doable tasks have to be done in your solution?
(b) Try to find a method that finds a better solution. (The "better" solution need not be an optimal solution.) Make a list of the employees with their non-doable tasks. What is the rationale behind your solution?
(c) Determine an optimal solution to this problem? Make a list of the employees with their non-doable tasks.

The preferences given above are not quite realistic. Employees are likely to have more detailed preferences. Assume that each employee makes a preference list of 1 through 10, where the task with label 1 is a very bad choice and the task with label 10 a very good choice. All preferences are listed in Table 4.7.

(d) Use the preferences from Table 4.7 to determine an assignment of employees to tasks with the highest sum of the levels of preference. Analyze the difference between the solution to this problem and the solution to part (c)?

It turns out that the excursion can be extended with one more day. It is decided that no more than three tasks and no less than two tasks are done by one person.

(e) If the organizing committee would have decided to extend the excursion for one more day before the excursion started, then what is an optimal assignment of employees to tasks?
(f) What would be an optimal solution if the decision was made on the Sunday during the excursion? Analyze the difference with part (e).
(g) Due to illness, employees 15, 29, and 35 are unable to join the excursion. This means that six tasks have to be done by the other colleagues. Again, no more than three tasks and no less than two tasks are done by one person. What is an optimal assignment of employees to tasks in case of part (d).
(h) Determine the tolerance interval of the coefficient representing the level of preference of employee 2 on task 6.

Problem 4.4. The Weakest Link
Within five years GTC will be using a new communication network with new phone devices. Making a mobile phone call seems easy these days, but six consecutive stages are needed to make the call possible. These stages are:

S1   calling the network;
S2   placing call into slot;
S3   recognizing the dialed number;
S4   finding nearest station to mobile phone of receiver;
S5   connecting to receiver; and
S6   connecting caller and receiver.

Table 4.7. Employees' preference ranking of the tasks

	Task							
Employee	1	2	3	4	5	6	7	8
1	7	8	10	2	10	9	9	4
2	10	10	2	8	3	8	1	5
3	8	1	9	7	7	10	1	5
4	1	10	3	5	8	9	4	7
5	5	2	9	9	8	9	10	6
6	9	7	10	3	7	6	1	8
7	8	2	10	10	5	8	2	9
8	8	10	10	3	6	8	3	6
9	4	3	5	6	9	4	10	1
10	1	6	3	7	7	1	5	8
11	4	7	9	3	2	10	10	10
12	5	10	9	6	4	2	2	7
13	7	5	8	8	4	7	4	10
14	3	10	1	8	4	5	4	2
15	9	2	8	2	1	10	10	4
16	7	1	5	10	6	3	3	7
17	5	4	7	10	3	4	5	2
18	9	7	2	6	4	10	10	1
19	3	3	9	10	5	7	7	10
20	2	9	10	5	3	1	2	9
21	7	3	1	2	10	10	6	9
22	9	1	6	3	3	4	3	5
23	6	5	3	2	6	10	1	2
24	10	3	10	1	1	7	6	7
25	5	2	10	5	10	3	1	6
26	1	9	2	10	4	4	1	8
27	9	4	2	1	10	10	4	5
28	2	6	8	5	4	10	2	9
29	9	2	1	6	3	10	10	9
30	8	3	8	9	8	10	10	5
31	9	2	7	3	2	10	10	5
32	8	2	1	5	8	6	7	6
33	4	1	9	10	3	2	1	10
34	5	6	2	7	3	4	2	6
35	4	7	10	7	6	2	6	6
36	1	8	4	5	10	4	1	3
37	4	2	6	9	10	1	5	5
38	8	7	4	6	6	8	10	1
39	4	7	5	1	6	7	6	9
40	3	6	7	10	9	6	2	3

In the market there are nine companies (labeled A through I), each producing six different devices, (labeled D1 through D6). For example, device 1 (D1) satisfies the requirements of stage 1 (S1). The devices all have different call transmitting times. The transmitting times are listed in Table 4.8. For example, D1 of company A needs 0.95100 seconds to transmit a call in S1. GTC does not want to more than one device

Table 4.8. Transmitting times for devices

Company	Device					
	D1	D2	D3	D4	D5	D6
A	0.95100	0.66100	0.92000	0.99890	0.66690	0.68490
B	0.69770	0.88410	0.85230	1.01810	0.79240	1.08400
C	1.31090	1.11730	1.00390	0.76340	0.35600	0.96130
D	1.13530	0.97130	1.32930	1.28780	0.29470	0.62630
E	0.28420	0.85810	1.00090	1.28850	1.05550	1.04390
F	0.32220	1.27000	1.23750	0.89210	0.50890	1.19070
G	1.28750	0.30920	0.83870	0.55110	1.17370	0.59930
H	0.77810	0.70780	0.84870	0.90820	0.35460	1.21800
I	0.96700	0.45090	0.53030	0.91880	0.82050	0.29540

from any single company. The goal of GTC is to assign devices to stages in such a way that the communication network is reliable and the total transmitting time of a call is as short as possible.

(a) By inspection, from which companies do you think GTC should purchase devices? How should GTC assign them to the stages? Determine the total transmitting time of your solution.

(b) From which companies should GTC purchase the devices to minimize transmission times? How should GTC assign them to the stages? Determine the total transmitting time.

(c) Determine the tolerance interval of the coefficient representing the transmitting time of D4 of company C in S4.

In part (b) we assumed the transmitting times to be deterministic. In fact, the transmitting times per device are normally distributed with means as given in Table 4.8, and standard deviation as given in Table 4.9. GTC guarantees its customers a 99% probability that a call is transmitted in less than three seconds.

(d) Does your solution in part (b) satisfy this guarantee? If not, find a solution that minimizes the total transmitting time but also satisfies this guarantee.

In an interview with its customers, GTC made an interesting observation. Until now, it was assumed that the customers are interested in the transmitting times of their calls. Instead, the customers complained more about the fact of being shut off due to overloaded lines. GTC has decided to use the capacity of the system as effective as possible. Therefore, it needs to know the limitations on the number of

Table 4.9. Standard deviations of transmitting times

Company	D1	D2	D3	D4	D5	D6
			Device			
A	0.03013	0.02427	0.04881	0.00370	0.04842	0.04261
B	0.01453	0.06145	0.02961	0.03979	0.01851	0.03469
C	0.02706	0.05837	0.03447	0.04122	0.02147	0.06122
D	0.05353	0.04073	0.07583	0.03955	0.03618	0.03329
E	0.01809	0.04413	0.06863	0.04681	0.06395	0.04285
F	0.02875	0.04750	0.04090	0.05767	0.02632	0.07661
G	0.03546	0.03084	0.04578	0.05175	0.04958	0.02862
H	0.02820	0.05516	0.05975	0.04661	0.03967	0.04175
I	0.03008	0.01325	0.03676	0.02310	0.05957	0.03281

simultaneous calls of each device. These limitations are shown in Table 4.10. For example, D1 of company A can process 106000 simultaneous calls in S1.

Table 4.10. Maximum number of simultaneous calls for each device

Company	D1	D2	D3	D4	D5	D6
			Device			
A	106	134	143	116	150	76
B	105	139	28	110	150	82
C	170	165	168	138	150	94
D	164	160	81	131	123	153
E	157	114	40	145	111	114
F	26	165	66	145	189	122
G	164	22	168	48	108	28
H	123	32	88	86	72	132
I	91	165	156	87	127	153

A call is transmitted when all devices have enough space to process the call. When there is one device fully occupied, the call is not transmitted. GTC wants to maximize the number of simultaneous calls that can be transmitted through the stages, but they still want to do this at minimum total transmitting time.

(e) Set up a model to solve this problem. What kind of matching problem is this?
(f) Give an optimal solution to the problem in part (e), and analyze the result.

Problem 4.5. Tie-in Sale Marketing Action

Once a year GTC offers a special sale. This year, one is thinking of a "tie-in" sale action. A tie-in sale is a sale where the customer can obtain the desired good (tying good) only if he/she agrees to purchase a different good (tied good). It is decided that the tie-in sale should include two products from two separate groups of purchasable articles. The customer has to choose one of ten different cell phones, and gets one other product for free. The cell phones and the other products are listed in Table 4.11.

Table 4.11. Equipment types

Cell phone type	Other product
GT I	Call bundle (100 min.)
GT II	Call bundle (200 min.)
GT III	Call bundle (300 min.)
GT IV	Call bundle (400 min.)
GT V	Extra Li-Ion Battery
GT VI	LED Flashing Keypad
GT VII	In-Car Charger
GT VIII	Belt-Clip Holster
GT IX	Handsfree Headset
GT X	USB Data Cable

So, GTC sells a cell phone together with one of the ten other products. The ten cell phones are the tying goods, and the other products are the tied goods. Since the other products stay, of course, available for normal sale, it is assumed that the sales of the other products are not influenced by this action.

Based on previous actions, GTC has made an estimation of the short-term sales increase of the phones, when combined with an other product. These estimates are given in Table 4.12. For example, GT I combined with Belt-Clip Holster, would yield an estimated short-term sales increase of GT I of 31%.

Table 4.12. Estimates of short-term sales increase of equipment (in %)

	GT									
	I	II	III	IV	V	VI	VII	VIII	IX	X
Call bundle (100 min.)	15	9	11	9	5	10	7	7	8	13
Call bundle (200 min.)	18	10	13	7	16	11	15	10	7	5
Call bundle (300 min.)	18	11	14	18	9	11	10	11	7	15
Call bundle (400 min.)	18	15	15	19	13	15	13	9	10	10
Extra Li-Ion Battery	20	20	18	22	18	20	17	19	13	15
LED Flashing Keypad	28	20	18	25	21	13	16	21	13	19
In-Car Charger	28	24	30	19	19	21	24	13	16	18
Belt-Clip Holster	31	34	31	29	23	24	21	21	23	16
Handsfree Headset	35	34	34	30	30	22	27	21	17	16
USB Data Cable	43	38	38	32	33	24	24	27	23	17

(a) Determine the minimum required short-term sales increase that makes the combination GT I with Extra Li-Ion Battery profitable, in terms of percentage. The purchase prices and selling prices are listed in Table 4.13.

The problem for GTC is to combine cell phones with other products, one cell phone to one other product, and vice versa, such as to maximize total profit.

Table 4.13. Purchase and selling price of equipment (in €)

Product	Purchase price	Selling price
Cell phone type		
GT I	100	300
GT II	110	330
GT III	120	360
GT IV	125	400
GT V	150	425
GT VI	160	480
GT VII	175	500
GT VIII	200	550
GT IX	250	600
GT X	300	700
Other product		
Call bundle (100 min.)	10	25
Call bundle (200 min.)	15	40
Call bundle (300 min.)	20	50
Call bundle (400 min.)	25	60
Extra Li-Ion Battery	30	75
LED Flashing Keypad	35	90
In-Car Charger	40	100
Belt-Clip Holster	45	125
Handsfree Headset	50	150
USB DataCable	55	160

(b) By inspection, how would you combine the products? What is the rationale behind your solution procedure?
(c) Set up a model to solve the problem. Determine and analyze your optimal solution.
(d) For the model used in part (c), draw and analyze the perturbation function of the coefficient representing the additional net sales of the combination GT IV with Call Bundle of 400 minutes.
(e) GTC discovered a mistake in the estimated short-term sales increase of phones combined with an Extra Li-Ion Battery. The estimated short-term sales increase of these combinations should be as shown in Table 4.14. Analyze the differences and determine the cost of the mistake.

Table 4.16. Skill and competency levels for each player

Qualities	Bryan	Joe	John	Jack	Michael	George	Bill	William	Phil	Simon	Matthew	Jason	Simon	Arnold	Pete	Sam	Jerri	Tom	Brad	Sean	Roger
1 Goalkeeping	3	2	5	9	7	9	8	6	5	5	5	4	4	8	4	7	4	7	8	5	3
2 Marking	7	5	4	9	5	6	7	6	7	7	7	3	2	2	1	1	2	2	9	3	3
3 Duel one-on-one	4	7	7	6	1	8	9	6	7	2	2	4	5	3	8	7	5	6	6	6	6
4 Passing over	7	7	5	3	9	6	5	9	6	7	7	7	2	7	5	8	4	6	7	2	5
5 Dribbling	7	2	2	7	9	3	2	2	5	3	3	6	5	3	4	7	5	5	5	8	7
6 Assist	5	8	8	5	2	3	3	8	7	7	7	1	4	6	8	1	6	1	9	5	6
7 Playing on the ball	3	6	7	2	3	4	4	1	7	8	8	8	6	5	5	6	2	4	6	2	5
8 Scoring	5	3	4	3	4	1	1	5	2	8	8	2	7	3	5	5	7	4	4	9	9
9 Heading	1	1	2	3	2	3	7	6	4	3	3	8	4	2	3	3	5	6	1	4	2
10 Short pass	4	4	2	8	4	7	5	8	4	1	1	8	2	7	6	3	7	3	4	7	8
11 Long pass	3	5	6	7	8	3	8	9	7	2	2	2	4	1	6	3	7	5	5	7	1
12 Shooting	7	2	1	9	1	6	6	6	1	4	4	2	7	2	2	1	5	3	1	1	1
13 Speed with ball	1	1	8	6	9	3	1	7	5	1	1	3	7	8	9	3	8	7	2	5	4
14 Right-foot play	5	3	5	4	8	2	6	3	6	1	1	5	1	3	9	3	3	9	1	6	7
15 Left-foot play	7	7	9	2	8	3	2	8	6	5	5	7	9	3	7	6	9	4	3	4	5
16 Speed off the ball	2	7	2	2	6	5	5	9	2	2	2	7	3	8	6	7	8	9	8	6	1
17 Strength	6	7	2	7	2	2	6	7	2	5	5	2	7	1	2	8	8	8	1	7	5
18 Coping with pressure	9	6	4	7	5	2	8	3	7	7	7	4	5	1	3	2	7	6	8	3	6
19 Coaching	5	8	8	6	3	5	4	3	7	6	6	8	4	3	3	5	4	4	1	5	6
20 Bringing order in play	4	7	9	4	6	7	8	7	5	2	2	1	2	2	6	2	8	2	9	2	7
21 Reading of the game	7	8	8	4	7	4	6	4	6	9	9	3	9	7	3	4	8	6	6	4	7
22 Positioning	5	7	4	7	4	5	4	6	6	6	6	7	8	4	4	2	2	9	3	6	4
23 Purposiveness	5	7	8	7	3	7	8	8	4	7	7	7	6	8	6	9	3	8	3	7	3
24 Creativity	4	8	5	1	3	3	6	3	8	4	4	3	7	9	4	3	1	9	2	7	7
25 Perseverance	1	2	8	4	3	8	3	8	2	9	9	4	2	6	9	1	8	7	4	9	3
26 Devotion	3	2	6	7	2	9	7	4	8	6	6	2	8	6	1	3	7	8	4	7	2
27 Consistency	4	4	8	6	3	5	2	2	8	4	4	3	1	8	5	7	5	6	4	6	4
28 Self-confidence	5	4	7	7	7	1	7	8	7	4	4	3	5	5	7	5	4	7	6	2	2
29 Team discipline	2	7	4	9	2	6	4	6	1	2	2	8	7	7	7	5	7	2	7	3	2
30 Value for the public	7	6	9	7	5	5	5	4	7	4	4	3	6	8	2	7	4	8	2	5	7

(h) Matthew wants to play on the position of Left Midfielder. How much has Matthew to improve his "skill on the ball" to become Left Midfielder in the starting line-up?

Problem 4.7. Company Expansion

In the early days, GTC was only active in the market of conventional telephone services. Nowadays, GTC is also active in the market of mobile telephony and Internet. Because of the increasing number of active businesses, the board of directors has decided to expand the number of jobs at five departments. To that end, GTC has advertised in various news media to acquire applicants to the new jobs. The number of new jobs for each department is given in Table 4.17. For instance, department 4 (D4) will have six new jobs. The contents of all jobs at a department are assumed to be equal. It turned out that GTC has received 35 job applications. The applicants are not

Table 4.17. New jobs in various departments

Department	1 2 3 4 5	Total
Number of new jobs	2 2 4 6 6	20

equally interested in each department and of course each department is not equally interested in each applicant. Each applicant ranks the departments from 1 through 5, where 1 is the best choice and 5 the worst. After several assessment sessions, each department ranks the applicants from 1 through 35, where 1 is the best choice and 35 is the worst. The preference levels of the applicants and the departments are listed in Table 4.18. In this table, the first number is the applicant's ranking of a department and the second number is the department's ranking of the applicant. So, the first choice of applicant 1 (A1) is D3. But this applicant is only twenty second on the ranking list of D3; the pair (1,22) is called the preference rating between A1 and D3.

(a) By inspection, how do you think GTC should assign the applicants to the departments? Explain the procedure used.
(b) GTC wants to assign applicants to departments in such a way that the total preference is optimum. What is an optimal solution to this problem?
(c) Give an assignment of applicants to departments that is optimal for the applicants.

In one of the possible solutions of part (b), A29 is assigned to a new job at D3. For this applicant it is suboptimal, since he ranked this department second among all departments. Another feature in this solution is the assignment of A18 to a new job at D5, although he was the ninth choice of D5. It could happen, that after the assignment is made, A29 tries to get the job at D5, because he prefers D5 to his current department. A29 is the eighth choice of D5. So, D5 prefers A29 to A18. D5 will fire A18 and hire A29.

(d) Find an assignment of applicants to departments that overcomes this problem.
(e) For several reasons A17 and A20 have decided to quit the selection procedure. Give an assignment of the remaining applicants to departments. Analyze the possible differences with the solution to part (d)?
(f) GTC has decided to create two more jobs in D3 and four more jobs in D5, in addition to the provisions in Table 4.17. Provide an assignment of applicants to departments in this situation.

Table 4.18. Preference ratings of applicants and departments

	D1	D2	D3	D4	D5
A1	(4,24)	(5,7)	(1,22)	(2,19)	(3,15)
A2	(1,31)	(5,20)	(2,29)	(3,15)	(4,22)
A3	(3,8)	(1,15)	(4,21)	(2,13)	(5,23)
A4	(3,21)	(4,16)	(1,15)	(2,6)	(5,13)
A5	(4,25)	(3,4)	(1,24)	(2,26)	(5,19)
A6	(3,15)	(2,33)	(4,18)	(5,1)	(1,12)
A7	(3,27)	(2,18)	(4,17)	(1,33)	(5,26)
A8	(3,30)	(5,14)	(2,27)	(4,17)	(1,31)
A9	(5,26)	(1,34)	(3,25)	(4,28)	(2,18)
A10	(1,2)	(3,32)	(5,2)	(2,7)	(4,5)
A11	(4,29)	(3,22)	(5,34)	(2,31)	(1,35)
A12	(1,22)	(5,9)	(3,5)	(4,16)	(2,24)
A13	(5,16)	(3,1)	(2,32)	(1,24)	(4,6)
A14	(2,19)	(1,13)	(4,35)	(3,34)	(5,2)
A15	(3,10)	(2,26)	(1,8)	(4,10)	(5,14)
A16	(2,33)	(1,11)	(3,28)	(4,25)	(5,34)
A17	(3,7)	(4,3)	(5,33)	(2,4)	(1,27)
A18	(1,35)	(5,35)	(3,13)	(4,11)	(2,9)
A19	(3,4)	(5,19)	(1,7)	(4,2)	(2,16)
A20	(3,28)	(4,25)	(1,4)	(5,14)	(2,33)
A21	(2,3)	(3,29)	(5,31)	(4,5)	(1,11)
A22	(5,11)	(4,31)	(2,10)	(1,27)	(3,32)
A23	(5,14)	(1,5)	(4,26)	(3,9)	(2,10)
A24	(1,32)	(4,8)	(3,9)	(5,30)	(2,25)
A25	(5,6)	(2,17)	(1,6)	(4,32)	(3,7)
A26	(5,12)	(3,21)	(1,1)	(2,3)	(4,21)
A27	(1,23)	(2,12)	(5,20)	(4,20)	(3,20)
A28	(4,9)	(1,30)	(2,11)	(3,8)	(5,4)
A29	(3,5)	(5,28)	(2,3)	(4,23)	(1,8)
A30	(4,20)	(2,10)	(5,19)	(1,12)	(3,28)
A31	(5,18)	(1,6)	(3,14)	(4,18)	(2,17)
A32	(1,13)	(3,2)	(2,23)	(4,35)	(5,1)
A33	(1,34)	(3,24)	(4,12)	(5,29)	(2,30)
A34	(2,17)	(3,23)	(5,16)	(4,21)	(1,29)
A35	(1,1)	(2,27)	(3,30)	(4,22)	(5,3)

5

Facility Location

5.1 Introduction

GTC has undertaken a major project to lay cables in a large region. The cables are supplied in spools. These spools are stored in adequate numbers in warehouses and transported to six cable laying sites. There are five sites for warehouses, labeled A, B, C, D, and E, at which GTC can rent warehouse space. Each of these sites have capacity sufficient to store all the spools required for the full duration of the project. The rents at different warehouse sites however, depend on the location of the site, and are given in Table 5.1. The rents quoted in Table 5.1 are the rents for a sufficiently large warehouse at that site for the full duration of the project. The cable laying sites

Table 5.1. Rents for warehouses at the different sites

	A	B	C	D	E
Rent (in €)	1,250,000	1,500,000	1,000,000	500,000	750,000

are labeled I, II, III, IV, V, and VI. The distances in kilometers to the cable laying sites from the potential warehouse sites along the existing road network are also known and are shown in Table 5.2. The transportation costs can be assumed to be €100 per kilometer for the full duration of the project. The demands for cables at the six sites I through VI are 700, 2300, 500, 3000, 2000, and 1500 spools, respectively.

GTC is faced with the problem of deciding where to rent warehouses, and which of the cable laying locations to supply from which warehouse, in an effort to minimize the total logistical costs. In this case, the logistical costs have two components, the costs for renting warehouses, and the costs of transporting cable spools from the warehouses to the cable laying locations. For example, if GTC decides to rent warehouses at locations B and C, then the warehouse at B would satisfy the cable requirements of cable laying sites I and VI, while the warehouse at C would satisfy the cable requirements of cable laying sites II, III, IV, and V. For this solution,

G. Sierksma and D. Ghosh, *Networks in Action: Text and Computer Exercises in Network Optimization*, International Series in Operations Research & Management Science 140, DOI 10.1007/978-1-4419-5513-5_8, © Springer Science + Business Media, LLC 2010

Table 5.2. Distances (in kms) from warehouse sites to cable laying sites

	I	II	III	IV	V	VI
A	2.9	2.6	4.8	4.6	3.7	2.5
B	2.3	4.8	4.4	4.4	3.4	3.3
C	2.3	2.4	3.8	2.9	3.3	4.7
D	3.1	2.6	4.3	3.5	4.6	2.8
E	3.6	2.1	4.1	4.8	2.8	2.8

GTC incurs €2,500,000 in rent and €$(230 \times 700 + 240 \times 2300 + 380 \times 500 + 290 \times 3000 + 330 \times 2000 + 330 \times 1500)$ = €2,928,000 in transportation costs, i.e., a total logistical cost of €5,428,000. GTC's problem of reducing logistical cost under these conditions is commonly referred to as the *uncapacitated facility location problem*, also called the *simple facility location problem*.

A slightly different problem arises when the warehouses at each of the locations are not sufficiently large. GTC then knows the capacity of the warehouses available at each potential site. Consider for example that the capacities of warehouses at different sites are as given in Table 5.3. The problem of deciding where to set up

Table 5.3. Capacities of warehouses at the various sites

	A	B	C	D	E
Capacity (in spools)	3000	5500	1500	2500	2700

warehouses, and how to supply cable location sites from the warehouses set up in presence of warehouse capacity constraints is called the *capacitated facility location problem*.

This problem is more complicated than the uncapacitated facility location problem because of the following reasons. Notice that all combinations of warehouses are not feasible solutions to the capacitated problem. For example, the solution considered in the uncapacitated case of renting warehouses only at B and C is no longer feasible, since the total capacity of these warehouses is only 8500 spools, while the total requirement for all the cable laying sites is 10,000 spools. Further, in an unca pacitated facility location problem, once a warehouse is rented, it is surely going to satisfy *all* the requirements of the cable laying sites for which it is the closest warehouse. Under capacity restrictions, a nearest warehouse may not be able to supply all the requirements of cable laying sites due to capacity limits. In other words, splitting of supply can occur in case of capacitated facility location problems, but not in uncapacitated facility location problems. Consider for example GTC's option of renting warehouses at sites A, B, and C. Their total capacity is 11,000 spools, and therefore they could supply the requirements of all the six cable laying sites. The total price that GTC would pay for renting these warehouses is €3,750,000. The requirements of the six cable laying stations would be met in the manner shown in Table 5.4 at a total transportation cost of €3,435,000. The total logistical cost for GTC in this

Table 5.4. Transportation plan when GTC rents warehouses at A, B, and C

	I	II	III	IV	V	VI
A		1000	500			1500
B	700	1300		1500	2000	
C				1500		

situation would therefore be €8,185,000. As we will see later, this is not the optimal solution for GTC.

We next describe two more location problems that complete the set of four problems that are most commonly studied as facility location problems.

Suppose that the research department of GTC is located in seven different buildings, labeled A, B, C, D, E, F, and G, in different parts of the city. These buildings house 12, 17, 15, 40, 30, 25, and 18 researchers, respectively. They all need access to certain library facilities. GTC is planning to build library facilities in two of the seven buildings and would like all researchers to access one of these facilities. Of course to make this decision it needs to know the distance between each pair of buildings along the road network in the city. Let us suppose that these distances are given in Table 5.5.

Table 5.5. Distances between different buildings

	A	B	C	D	E	F	G
A	—	5.9	5.7	4.7	3.8	8.5	4.6
B		—	4.0	2.9	2.9	3.4	1.4
C			—	1.1	6.9	5.7	5.4
D				—	5.8	6.3	4.3
E					—	6.3	4.3
F						—	4.4

There are two main ways in which GTC can take the decision about where to locate the two library facilities.

One way to distribute library facilities equitably would be to minimize the sum of the distances that researchers need to travel to reach library facilities. If the number of researchers in the buildings are not taken into consideration, then the problem is referred to as the *2-median problem*. If the distances however are weighted, then we have the *weighted 2-median problem*. A logical set of weights in this case would be proportional to the number of researchers working in the building. Suppose that GTC decides to set up library facilities in buildings B and C. Then the library in building B would cater to researchers in building B, E, F, and G, while the library in building C would cater to researchers in A, C, and D. For the unweighted case, this solution would have the sum of distances as $5.7 + 0.0 + 0.0 + 1.1 + 2.9 + 3.4 + 1.4$ = 14.5 kilometers, while if the distances are weighted by the number of researchers in each building, the sum would be $12 \times 5.7 + 17 \times 0.0 + 15 \times 0.0 + 40 \times 1.1 + 30 \times$ $2.9 + 25 \times 3.4 + 18 \times 1.4 = 309.6$ kilometers. GTC's objective in 2-median problems is to locate libraries in buildings that minimize these values.

Another way is to ensure that the maximum distance that researchers need to travel to reach the nearest library facility is as small as possible. If the numbers of researchers in the buildings are not taken into consideration, then the problem is referred to as the *2-center problem*. If however, we want to weigh the distance that researchers from any building need to travel, possibly with weights corresponding to the number of researchers in a particular building, then the problem is known as the *weighted 2-center problem*. If GTC decides to set up library facilities in buildings B and C, then the maximum distance that any researcher would have to travel to use library facilities is $\max\{5.7, 0.0, 0.0, 1.1, 2.9, 3.4, 1.4\} = 5.7$ kilometers in the unweighted case, and $\max\{12 \times 5.7, 17 \times 0.0, 15 \times 0.0, 40 \times 1.1, 30 \times 2.9, 25 \times 3.4, 18 \times 1.4\} = 87$ kilometers if the distances are weighted by the number of researchers in each building. GTC's objective in 2-center problems is to locate libraries in buildings that minimize these values.

Of course, the problems described above can be generalized into (weighted) p-median and (weighted) p-center problems, if we deal with situations where p facilities (with $p \geq 1$) need to be located. The solution to the p-median and the p-center problems are called the *p-median* and *p-center* of the network respectively.

We have introduced four classes of location problems in this section. Table 5.6 provides the characteristics of each of the problems.

Table 5.6. The four location problems

	Inter-client distances	Facility setup costs	Facility capacities	Objective
Uncapacitated location	present	present	absent	minimize sum of costs
Capacitated location	present	present	present	minimize sum of costs
p-Median	present	absent	absent	minimize sum of distances
p-Center	present	absent	absent	minimize the maximum distance

An important extension of p-median and p center problems is one in which the medians or centers need not be located at nodes of the network, but can also be located at points along the arcs of the network. These extensions are called the *absolute p-median* and *absolute p-center* problems respectively.

Consider for example, the network shown in Figure 5.1. The numbers next to the edges of the network represent the distance between the two nodes on which the edge is incident. Suppose that we want to locate two facilities on this network

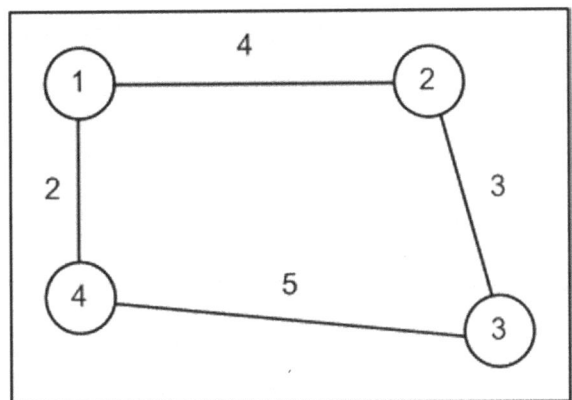

Fig. 5.1. Network to illustrate the absolute p-center problem

such that the maximum distance between a node and its nearest facility is as small as possible. If we solve the problem as a p-center problem, then an optimal location of the centers will be at nodes 1 and 2, for which the maximum distance between a node and its nearest facility will be 3 units. If however, we solve it as an absolute p-center problem, then the optimal location of the two facilities will be the midpoints of edge 1–4 and 2–3, in which case the maximum distance between a node and its nearest facility reduces to 2.5 units.

Obviously, the objective function value of an optimal solution to the absolute p-median (or absolute p-center) problem does not exceed the optimal objective function value for the p-median (respectively, p-center) problem defined on the same network. However, it can be shown that when $p = 1$, an optimal solution to the p-median (or p-center) problem is also an optimal solution to the absolute p-median (respectively, absolute p-center) problem defined on the same network.

5.2 Applications

Facility location problems form the basis of many practical optimization problems. Most uncapacitated and capacitated facility location problems are actually warehouse location problems, exactly like the ones described in Section 5.1. However, applications of p-median and p-center problems are often more subtle. Two of these applications are sketched below.

5.2.1 Cluster analysis

Data clustering is a common technique in many statistical data analysis methods used in many fields such as bioinformatics, psychology, medicine, economics, and marketing. The objective of cluster analysis is to group data or variables into collections which have a high degree of natural association, such that such association among elements of different clusters is low. In these applications, one uses a pre-determined number of clusters, say p, and defines "distances" between data points as a function of their natural association levels. Then using all the data points as candidate medians, one solves a p-median problem maximizing the sum of distances to obtain the clusters.

5.2.2 Locating undesirable facilities

The location of garbage depots in a location has social costs, since it can lower the quality of life for the people living nearby. Such social costs reduce with distance of a location from the depot. However, the garbage depots need to be located such that residents can access them without traveling long distances. The location of such garbage depots are often modeled as restricted p-center problems. The number and possible locations of garbage depots are first obtained. Then using social costs to measure distances between population locations and possible garbage sites, a p-center problem is solved with additional restrictions on the maximum physical distances between population locations and garbage depots.

5.3 Linear Programming Formulations

Facility location problems can be solved to optimality using linear programming. In this section we provide linear programming formulations to the four location problems described in the last section.

5.3.1 The uncapacitated facility location problem

For an uncapacitated facility location problem, we require two types of variables. The first type consists of binary variables y_i, one for each warehouse site. y_i attains a value of 1 if we decide to rent a warehouse at site i, and 0 otherwise. In addition to these variables, we have continuous non-negative variables x_{ij} modeling the number of spools that are transported from the warehouse at site i to the cable laying site at site j in a solution to the problem.

The objective is to minimize the total logistical cost, i.e.,

$$\text{Minimize} \sum_i r_i y_i + \sum_i \sum_j t_{ij} x_{ij}, \tag{5.1}$$

where r_i is the rent of the warehouse at site i, and t_{ij} is the unit transportation cost from a warehouse at site i to a cable laying site at location j. The first term in the objective function is called the setup cost, and the second term in the function is called the transportation cost.

Two sets of constraints are sufficient to describe feasible solutions to the uncapacitated location problem. The constraints in the first set are called demand constraints, and they ensure that the demands of all the cable laying sites are satisfied. This constraint for cable laying site j takes the following form.

$$\sum_i x_{ij} = d_j, \tag{5.2}$$

where d_j is the demand of the cable laying site j.

The constraints in the other set are called supply constraints. They ensure that a warehouse site can supply spools to cable laying sites only if a warehouse has been rented at that site. The supply constraint for warehouse site i is of the following form.

$$\sum_j x_{ij} \leq My_i, \tag{5.3}$$

where M is a number larger than the sum of all demands. When $y_i = 1$ for any warehouse site i, the right-hand side of the corresponding supply constraint is sufficiently large, which makes the constraint redundant. However when $y_i = 0$, the right hand side reduces to 0, so that no x_{ij} can assume non-zero values.

A complete formulation for the uncapacitated facility location problem in which the set of warehouse locations is denoted by I, the set of cable laying sites is denoted by J, the rent vector is denoted by $R = (r_i)$, the demand vector by $D = (d_j)$, and the cost matrix is denoted by $T = [t_{ij}]$ is shown in Figure 5.2.

Minimize

$$z = \sum_{i \in I} r_i y_i + \sum_{i \in I} \sum_{j \in J} t_{ij} x_{ij}$$

Subject to

$$\sum_{i \in I} x_{ij} = d_j \quad \text{for all } j \in J$$

$$\sum_{j \in J} x_{ij} \leq My_i \quad \text{for all } i \in I$$

$$x_{ij} \geq 0 \quad \text{for each } i \in I, j \in J$$

$$y_i \in \{0, 1\} \text{ for each } i \in I$$

Fig. 5.2. Linear programming formulation of the uncapacitated facility location problem

As an illustration, we provide the linear programming formulation described in Section 5.1 using data in Tables 5.1 through 5.3 in Figure 5.3.

Minimize

$$z = 1250000y_A + 1500000y_B + \cdots + 750000y_E + 290x_{A,1} + 260x_{A,II} + \cdots + 280x_{E,V} + 280x_{E,VI}$$

Subject to

$$x_{A,I} + x_{B,I} + x_{C,I} + x_{D,I} + x_{E,I} = 700 \quad \text{(Constraint 5.2 for I)}$$
$$x_{A,II} + x_{B,II} + x_{C,II} + x_{D,II} + x_{E,II} = 2300 \quad \text{(Constraint 5.2 for II)}$$

There are four more similar constraints for III, IV, V, and VI.

$$x_{A,I} + x_{A,II} + x_{A,III} + x_{A,IV} + x_{A,V} + x_{A,VI} \leq My_A \quad \text{(Constraint 5.3 for A)}$$
$$x_{B,I} + x_{B,II} + x_{B,III} + x_{B,IV} + x_{B,V} + x_{B,VI} \leq My_A \quad \text{(Constraint 5.3 for B)}$$

There are three more similar constraints for C, D, and E.

$$y_A, y_B, y_C, y_D, y_E \in \{0, 1\} \quad \text{(Binary location variables)}$$
$$x_{A,I}, x_{A,II}, \ldots, x_{E,V}, x_{E,VI} \geq 0 \quad \text{(Non-negative flows)}$$

Fig. 5.3. Formulation of the uncapacitated facility location problem in Section 5.1

The optimal solution here is to set $y_D = 1$, $y_A = y_B = y_C = y_E = 0$, and to set all x_{ij}'s to 0 except $x_{D,I}$, $x_{D,II}$, $x_{D,III}$, $x_{D,IV}$, $x_{D,V}$, and $x_{D,VI}$, which are set to 700, 2300, 500, 3000, 2000, and 1500, respectively. This means that the optimal course of action for GTC would be to rent a warehouse only at location D, and to supply all the cable laying sites from this warehouse. The rent for this solution would be €500,000 and the total transportation cost would be €3,420,000, bringing the total logistical cost to €3,920,000.

5.3.2 The capacitated facility location problem

The capacitated location problem can be formulated as a simple extension of the uncapacitated location problem. The decision variables are identical to those used in the uncapacitated location problem described above; we have binary decision variables y_i which indicate whether or not a warehouse is to be rented at warehouse site i, and continuous non-negative variables x_{ij} denoting the number of spools transported from a warehouse at location i to a cable laying site at location j.

The objective function and the demand constraints are both identical to those in the formulation for the uncapacitated location problem (i.e., of the form (5.1) and (5.2), respectively). The supply constraints however are different, since they need to take care of the capacities of the warehouses.

One way of implementing the supply constraints is to have two sets of constraints. The first set, identical to the supply constraints in (5.3), ensure that no spool is supplied from a warehouse that has not been rented. The second set of constraints ensure that a warehouse does not supply more than its capacity. A typical constraint from this set, for site i is shown in (5.4), in which c_i refers to the capacity of a warehouse at site i.

$$\sum_j x_{ij} \le c_i, \tag{5.4}$$

The two sets of constraints are normally combined into one set in linear programming formulations for the capacitated location problem, namely,

$$\sum_j x_{ij} \le c_i y_i. \tag{5.5}$$

When $y_i = 1$ for any warehouse site i, the constraint reduces to the form $\sum_j x_{ij} \le c_i$ which is identical to the constraint (5.4). However, if $y_i = 0$ constraint (5.5) reduces to the form $\sum_j x_{ij} \le 0$, which ensures that no cable laying site is supplied by a warehouse at site i.

A complete formulation for the capacitated facility location problem in which the set of warehouse locations is denoted by I, the set of cable laying sites is denoted by J, the rent vector is denoted by $R = (r_i)$, the capacity vector by $C = (c_i)$, the demand vector by $D = (d_j)$, and the cost matrix is denoted by $T = [t_{ij}]$ is shown in Figure 5.4.

Minimize

$$z = \sum_{i \in I} r_i y_i + \sum_{i \in I} \sum_{j \in J} t_{ij} x_{ij}$$

Subject to

$$\sum_{i \in I} x_{ij} = d_j \quad \text{for all } j \in J$$

$$\sum_{j \in J} x_{ij} \le c_i y_i \quad \text{for all } i \in I$$

$$x_{ij} \ge 0 \quad \text{for each } i \in I, \; j \in J$$

$$y_i \in \{0, 1\} \text{ for each } i \in I$$

Fig. 5.4. Linear programming formulation of the capacitated facility location problem

As an illustration, we provide the linear programming formulation described in Section 5.1 in Figure 5.5.

The optimal solution to this problem is to set $y_B = y_D = y_E = 1$ and $y_A = y_C = 0$. The optimal values of the x_{ij} variables are given in Table 5.7, in which the number in the ith row and jth column denotes the value of x_{ij} in the optimal solution.

This means that the optimal course of action for GTC, given the capacities of the warehouses, and the rents, is to rent warehouses at B, D, and E. They should

Minimize

$$z = 1250000y_A + 1500000y_B + \cdots + 750000y_E + 290x_{A,1} + 260x_{A,II} + \cdots + 280x_{E,V} + 280x_{E,VI}$$

Subject to

$$x_{A,I} + x_{B,I} + x_{C,I} + x_{D,I} + x_{E,I} = 700 \quad \text{(Constraint 5.2 for I)}$$
$$x_{A,II} + x_{B,II} + x_{C,II} + x_{D,II} + x_{E,II} = 2300 \quad \text{(Constraint 5.2 for II)}$$

There are four more similar constraints for III, IV, V, and VI.

$$x_{A,I} + x_{A,II} + x_{A,III} + x_{A,IV} + x_{A,V} + x_{A,VI} \leq 3000y_A \quad \text{(Constraint 5.5 for A)}$$
$$x_{B,I} + x_{B,II} + x_{B,III} + x_{B,IV} + x_{B,V} + x_{B,VI} \leq 5500y_A \quad \text{(Constraint 5.5 for B)}$$

There are three more similar constraints for C, D, and E.

$$y_A, y_B, y_C, y_D, y_E \in \{0,1\} \quad \text{(Binary location variables)}$$
$$x_{A,I}, x_{A,II}, \ldots, x_{E,V}, x_{E,VI} \geq 0 \quad \text{(Non-negative flows)}$$

Fig. 5.5. Formulation of the capacitated facility location problem in Section 5.1

Table 5.7. The optimal solution to the capacitated location problem in Section 5.1

x_{ij}	I	II	III	IV	V	VI
A	0	0	0	0	0	0
B	700	0	500	500	1600	1500
C	0	0	0	0	0	0
D	0	0	0	2500	0	0
E	0	2300	0	0	400	0

then supply 2500 spools for the cable laying site at IV from the warehouse at D, the full requirement of cable laying site II and 400 spools for the cable laying site at V from the warehouse at E, and the remainder of the requirements from the warehouse at B. The rent component of this solution is €2,750,000 and the transportation cost component is €3,110,000. Therefore the least logistical cost that GTC would have to incur in this setup is €5,860,000.

5.3.3 The *p*-median problem

The linear programming formulation for the *p*-median problem requires binary decision variables only. Recall the *p*-median problem introduced in Section 5.1. Let *B* be the set of buildings and *L* be the set of libraries that need to be located in the buildings. We define decision variables of the form y_{ij}, each of which assumes a value of 1 if researchers at building *i* use the library facilities at building *j*. In this representation, any building i for which $y_{ii} = 1$ represents a building in which a facility should be located. We will assume that the distance between buildings i and j are denoted by d_{ij}. So d_{ii} is zero for all values of i. The number of people working in building i (which will be taken as a weight for the weighted version of the problem) is denoted by w_i.

In the unweighted version of the problem, the objective is to minimize the sum of the distances between the buildings and their nearest facility. So it is represented as follows.

$$\text{Minimize} \sum_{i \in B} \sum_{j \in L} d_{ij} y_{ij} \tag{5.6}$$

In the weighted version, each term of the objective function would be appropriately weighted, so that the objective is represented as follows.

$$\text{Minimize} \sum_{i \in B} \sum_{j \in L} (w_i d_{ij}) y_{ij} \tag{5.7}$$

Constraints in the p-median problem need to ensure several things. First, they need to ensure that a total of exactly p 'medians' ($p \geq 1$) are located. In terms of the problem described in Section 5.1, this will ensure that exactly two library facilities are set up, i.e., $p = 2$. The constraint that ensures this is

$$\sum_{i \in B} y_{ii} = p. \tag{5.8}$$

Next there need to be constraints that ensure that each point is assigned to exactly one facility. In terms of GTC's problem, these constraints would ensure that researchers from each building visit library facilities in one building only. For any building i, this constraint can be modeled as

$$\sum_{j \in L} y_{ij} = 1. \tag{5.9}$$

Finally, there need to be constraints that ensure that building i is assigned to a library at j only if there is a library at point j. An indicator of whether a facility is located at point j is the decision variable y_{jj}. In terms of GTC's problem, these constraints would ensure that researchers from a building visit library facilities in any building only if library facilities are set up in the latter. The constraint that ensures this is

$$y_{jj} - y_{ij} \geq 0. \tag{5.10}$$

The complete linear programming model for a weighted p-median problem is given in Figure 5.6.

As an illustration, we provide the linear programming formulation of the p-median problem described in Section 5.1 in Figure 5.7.

The optimal solution has y_{AD}, y_{BB}, y_{CD}, y_{DD}, y_{EB}, y_{FB}, and y_{GB} set to 1. The remaining variables are set to zero. This means that optimally, GTC should set up library facilities in buildings B and D. Researchers from buildings B, E, F, and G would use the library facilities at building B, while those from buildings A, C, and D would use the library facilities in building D. The total weighted distance traveled in this solution is 270.1 kilometers.

Minimize

$$z = \sum_{i \in B} \sum_{j \in L} (w_i d_{ij}) y_{ij}$$

Subject to

$$\sum_{i \in B} y_{ii} = p$$

$$\sum_{j \in L} y_{ij} = 1 \qquad \text{for all } i \in B$$

$$y_{jj} - y_{ij} \geq 0 \qquad \text{for all } i \in B, \ j \in L$$

$$y_{ij} \in \{0,1\} \text{ for each } i \in B, \ j \in L$$

Fig. 5.6. Linear programming formulation of the p-median problem

5.3.4 The p-center problem

The formulation of the p-center problem is related to the formulation of the p-median problem in almost the same way as the formulation of the capacitated location problem is related to the formulation of the uncapacitated location problem. We use the same decision variables, i.e., binary decision variables y_{ij} which assume a value of 1 if researchers from the building at i use the library facilities in the building at j. Also, as in the p-median problem, we conclude that library facilities are located in a building at i if and only if $y_{ii} = 1$. As before, d_{ij} represents the distance between buildings i and j, and w_i represents the number of people working in building i.

The constraints that ensure that exactly p facilities are located, that researchers in each building visit library facilities in one building only, and that library facilities can be used in one building only if they are set up there, are implemented exactly in the same manner as they are implemented in the p-median problem, namely by constraints (5.8), (5.9), and (5.10), respectively. The difference for this formulation is that here we have to minimize the maximum distance traveled by researchers.

In the unweighted problem, we formulate this in the following manner. We define a decision variable v, which stores the maximum distance that researchers from any building have to travel. Researchers from building i have to travel a distance of $\sum_j d_{ij} y_{ij}$ to the nearest building. So for each building i, we add a constraint of the form

$$v - \sum_j d_{ij} y_{ij} \geq 0, \tag{5.11}$$

and set the objective to

$$\text{Minimize } v. \tag{5.12}$$

In the weighted p-center problem we must minimize the maximum weighted distance that researchers need to travel. Here too, we define a decision variable v,

Minimize

$$z = 17 \times 5.9 y_{BA} + 15 \times 5.7 y_{CA} + \cdots + 18 \times 4.6 y_{GA} +$$
$$12 \times 5.9 y_{AB} + 15 \times 4.0 y_{CB} + \cdots + 18 \times 1.4 y_{GB} +$$
$$\cdots \qquad \cdots \qquad \cdots$$
$$18 \times 4.6 y_{AG} + 17 \times 1.4 y_{BG} + \cdots + 25 \times 4.4 y_{FG}$$

Subject to

$y_{AA} + y_{BB} + y_{CC} + y_{DD} + y_{EE} + y_{FF} + y_{GG} = 2$	(Constraint (5.8))
$y_{AA} + y_{AB} + y_{AC} + y_{AD} + y_{AE} + y_{AF} + y_{AG} = 1$	(Constraint (5.9) for A)
$y_{BA} + y_{BB} + y_{BC} + y_{BD} + y_{BE} + y_{BF} + y_{BG} = 1$	(Constraint (5.9) for B)

There are five more similar constraints for C, D, E, F, and G.

$y_{AA} - y_{BA} \geq 0$	(Constraint (5.10) from B to A)
$y_{AA} - y_{CA} \geq 0$	(Constraint (5.10) from C to A)
$y_{AA} - y_{DA} \geq 0$	(Constraint (5.10) from D to A)
$y_{AA} - y_{EA} \geq 0$	(Constraint (5.10) from E to A)
$y_{AA} - y_{FA} \geq 0$	(Constraint (5.10) from F to A)
$y_{AA} - y_{GA} \geq 0$	(Constraint (5.10) from G to A)

There are six more similar sets of constraints for B, C, D, E, F, and G.

$$y_{AA}, y_{AB}, y_{AC}, \ldots y_{GF}, y_{GG} \in \{0,1\} \quad \text{(Binary variables)}$$

Fig. 5.7. Formulation of the p-median problem in Section 5.1

which stores the maximum weighted distance that researchers from any building have to travel. The constraint for building i in this case that is equivalent to (5.11) is

$$\sum_j (w_{ij} d_{ij}) y_{ij} \leq v, \tag{5.13}$$

and the objective here too is to minimize the value of v.

The complete formulation of the p-center problem therefore is the one given in Figure 5.8.

As an illustration, we show the formulation of GTC's library facility location problem using the number of researchers in a building as the weights for the building in Figure 5.9.

The optimal solution has y_{AC}, y_{BB}, y_{CC}, y_{DC}, y_{EB}, y_{FB}, and y_{GB} set to 1. The remaining variables are set to 0. The value of v is 87. This means that optimally, GTC should set up library facilities in buildings B and C. Researchers from buildings B, E, F, and G would use the library facilities at building B, while those from

Minimize

$$z = v$$

Subject to

$$\sum_{i \in B} y_{ii} = p$$

$$\sum_{j \in L} y_{ij} = 1 \qquad \text{for all } i \in B$$

$$y_{jj} - y_{ij} \geq 0 \qquad \text{for all } i \in B,\ j \in L$$

$$v - \sum_{j} (w_{ij} d_{ij}) y_{ij} \geq 0 \qquad \text{for all } i \in B$$

$$y_{ij} \in \{0, 1\} \text{ for each } i \in B,\ j \in L$$

Fig. 5.8. Linear programming formulation of the p-center problem

buildings A, C, and D would use the library facilities in building D. The maximum weighted distance that researchers from any building have to travel is 87 kilometers corresponding to researchers from building E.

5.4 Algorithms for Location Problems

The location problems described in this chapter are of a different level of difficulty than most of the other problems that we have dealt with so far. They belong to a class commonly referred to as NP-hard. This means that, although some particular location problems may be easy to solve, nobody has come up with any algorithm that will solve all location problems to optimality in reasonable time. The term reasonable, when it is used in this context, has a precise definition; however the definition is not pertinent to our discussion.

Even though there are no algorithms that solve all location problems in reasonable time, there are algorithms (called exact algorithms) that solve such location problems to optimality, but may take very long execution times. There are also other algorithms, called heuristics that generate high quality solutions in very reasonable times. In this section, we will describe one exact algorithm and one heuristic for location problems. We will use the uncapacitated location problem to show how these work.

5.4.1 An exact algorithm: Branch and bound

The exact algorithm that we describe here is a general purpose algorithm called a *branch and bound algorithm*. As the name suggests, there are two main interwoven components of the algorithm, namely, branching and bounding.

Minimize

$$z = v$$

Subject to

$$y_{AA} + y_{BB} + y_{CC} + y_{DD} + y_{EE} + y_{FF} + y_{GG} = 2 \quad \text{(Constraint (5.8))}$$
$$y_{AA} + y_{AB} + y_{AC} + y_{AD} + y_{AE} + y_{AF} + y_{AG} = 1 \quad \text{(Constraint (5.9) for A)}$$
$$y_{BA} + y_{BB} + y_{BC} + y_{BD} + y_{BE} + y_{BF} + y_{BG} = 1 \quad \text{(Constraint (5.9) for B)}$$

There are five more similar constraints for C, D, E, F, and G.

$$y_{AA} - y_{BA} \geq 0 \quad \text{(Constraint (5.10) from B to A)}$$
$$y_{AA} - y_{CA} \geq 0 \quad \text{(Constraint (5.10) from C to A)}$$
$$y_{AA} - y_{DA} \geq 0 \quad \text{(Constraint (5.10) from D to A)}$$
$$y_{AA} - y_{EA} \geq 0 \quad \text{(Constraint (5.10) from E to A)}$$
$$y_{AA} - y_{FA} \geq 0 \quad \text{(Constraint (5.10) from F to A)}$$
$$y_{AA} - y_{GA} \geq 0 \quad \text{(Constraint (5.10) from G to A)}$$

There are six more similar sets of constraints for B, C, D, E, F, and G.

$$v - (17 \times 5.9 y_{BA} + 15 \times 5.7 y_{CA} + \cdots + 18 \times 4.6 y_{GA}) \geq 0 \quad \text{(Constraint (5.13) at A)}$$
$$v - (12 \times 5.9 y_{AB} + 15 \times 4.0 y_{CB} + \cdots + 18 \times 1.4 y_{GB}) \geq 0 \quad \text{(Constraint (5.13) at B)}$$

There are five more similar constraints for C, D, E, F, and G.

$$y_{AA}, y_{AB}, y_{AC}, \cdots y_{GF}, y_{GG} \in \{0,1\} \quad \text{(Binary variables)}$$

Fig. 5.9. Formulation of the p-center problem in Section 5.1

The branching component of the algorithm is a systematic way in which one can generate *all* feasible solutions to the problem. Given an uncapacitated location problem with k warehouse sites ($k \geq 1$), the branching process would systematically generate all $2^k - 1$ combinations of sites at which warehouses can be rented. For each combination, since we know the warehouses that are located, it is easy to compute which is the nearest warehouse to each cable laying site, and hence the total logistical cost for that combination. In theory, once we are able to evaluate all $2^k - 1$ combinations, we could choose the one among them with least cost, and that would be an optimal solution. Obviously, since $2^k - 1$ becomes uncontrollably large when k is even moderately large, the branching component by itself would make a very inefficient algorithm, which would be useless even for moderately sized problems.

We therefore use a bounding component in the algorithm. Bounds operate on subproblems. A subproblem is a problem obtained by adding constraints to the original problem. These constraints force us to look at only a subset of the feasible region of the original problem. Let us suppose, for example, that we create a subproblem by adding the constraint $y_B = 1$ to the uncapacitated location problem formulated in Figure 5.1. This subproblem restricts us to look at only those solutions to the uncapacitated location problem in which we decide to rent a warehouse at location B. If we look at a subproblem with the additional constraint $y_C = 0$, then we are looking only at those solutions for which we decide to rent a warehouse at location B, and to not rent a warehouse at location C. Given a subproblem, the bounding scheme gives us a bound, i.e., an optimistic estimate of the logistical cost that would be incurred in a best solution to that subproblem.

How do we use this bounding scheme? Suppose that we have a feasible solution at hand, and the logistical cost for this solution is €4,000,000. Now suppose that at a subproblem, we find a bound of €4,500,000. Since this is an optimistic estimate to the best solution for that subproblem, it is impossible that solving this subproblem will yield a solution that is better than the one we already have. So there is no point evaluating the subset of solutions to the original problem that are also solutions to this subproblem. If on the other hand, the bound turned out to be €3,800,000, then there is a possibility of getting a solution out of this subproblem that is better than the one we have at hand. In such a case, it makes sense to explore this subproblem further. This bounding method allows us to ignore a large subset of the $2^k - 1$ feasible solutions to the original problem, and arrive at an optimal solution more quickly. It should be obvious that a better bound, i.e., an estimate that is closer to the logistical cost of the best solution to the subproblem would help us to solve the problem quicker, if we do not spend too much additional effort computing the better bound.

How do we arrive at such bounds? There are many ways of computing bounds, but an easy way of getting a bound is to generate relaxations of the formulation of the uncapacitated location problem that are fast to solve, and then use the optimal objective function values of these relaxations. For example, if we replace the y_i binary variables in the formulation in Figure 5.2 with continuous variables which can take values between 0 and 1, we create a linear programming relaxation of the uncapacitated location problem. This problem is solved very quickly by linear programming solvers. The optimal objective function value from this relaxation can be used as a bound for the uncapacitated location problem.

We are now in a position to describe the branch and bound algorithm. It works by maintaining a list of subproblems generated, which we call LIST. At each iteration, the algorithm chooses a subproblem from LIST, removes it from there, and works on it. The algorithm first generates a bound for this subproblem. If the bound is worse than the cost of the best solution that the algorithm has found so far, then the iteration is over. If it is not, then the algorithm looks at the solution to the relaxation that provided the bound. This solution may be feasible for the original problem, or it may not be. If the solution was a feasible solution to the original problem, then the solution is better than the best solution the algorithm had found so far. In that case, the best solution found by the algorithm is updated with this solution, and the iteration is over. If the solution that provided the bound was not a feasible solution to the original problem, then the branching procedure is started. Typically, the algorithm chooses a component of the solution to the relaxation that caused the solution not to be feasible to the original problem, and creates more subproblems by adding constraints about this component. The subproblems that are formed should cover all possible solutions feasible to the original problem that were feasible to the subproblem from LIST taken out at the beginning of the iteration. These new subproblems are added to LIST, and the iteration is over. In case of the uncapacitated location problem, for example, if the bound is obtained using a linear programming relaxation, then a decision variable in the solution to the relaxation which has fractional values would be a component that causes the solution to the relaxation to be infeasible to the original problem. In that case, two subproblems may be formed by forcing the decision variable to attain values of 0 and 1. The algorithm starts by putting the original problem in LIST, and terminates when LIST is empty at the start of an iteration. The best solution found by the algorithm at that point is output as an optimal solution.

We now illustrate the branch and bound algorithm using the uncapacitated location problem described in Section 5.1. We set the value of M used in the formulation in Figure 5.2 to 10,001, since the sum of all requirements for cable laying stations was 10,000 spools, and use a linear programming relaxation to obtain the bounds. An optimal solution to the linear programming relaxation for a subproblem is indicated using the vector $\{y_A, y_B, y_C, y_D, y_E\}$ in which each of the entries in the vector lies within the interval $[0, 1]$. The variable BEST stores the best solution found so far, and is output on termination of the algorithm. The variable BESTCOST stores the logistical cost associated with BEST. Subproblems are recursively defined using the notation $P_k = \{P_m, y_j = r\}$, where P_k is a subproblem of P_m in which the additional constraint $y_j = r$ has been added.

At the beginning of the first iteration, LIST contains the original problem. No solution had been found to the problem so far, so BEST = $\emptyset$, and BESTCOST = ∞. During the first iteration the original problem P_0 is taken out of LIST and the bound for this problem is computed. The bound turns out to be €3,501,418.22 corresponding to $\{0, 0, 0.42, 0.15, 0.43\}$. Since this is not a feasible solution to the original problem, we need to perform a branching operation. We choose y_C as the component to branch on. So two subproblems are formed: $P_1 = \{P_0, y_C = 1\}$, and $P_2 = \{P_0, y_C = 0\}$. These two subproblems are added to LIST and the first iteration is over. The second iteration begins with LIST = $\{P_1, P_2\}$, BEST = $\emptyset$, and BESTCOST = ∞. The bound for P_1 is €3,927,992.49, and that for P_2 is €3,552,439.26. Since the bound for P_2 is lower, we choose to use it in the second iteration. The solution corresponding to the bound is $\{0, 0, 0, 0.57, 0.43\}$. Since this solution is not feasible for the original subproblem, we invoke the branching operation again. We branch on y_D and create two subproblems, namely $P_3 = \{P_2, y_D = 1\}$, and $P_4 = \{P_2, y_D = 0\}$, which we add to LIST before terminating the iteration. The branch and bound algorithm terminates at the eighth iteration. The details of all the iterations are given in Table 5.8.

5.4.2 A non-exact algorithm: Greedy heuristic

Many heuristic algorithms have been formulated to solve location problems. In this part, we describe a greedy heuristic. A greedy heuristic is one that is myopic, or short-sighted, in the sense that at any decision point, it takes that decision which looks most

Table 5.8. Branch and bound in action

Iteration	LIST	BEST	BESTCOST (in €)	Sub-problem chosen	Bound (in €)	Decision
1	$\{P_0\}$	$\emptyset$	∞	P_0	3,501,418.22	Branch on y_C
2	$\{P_1, P_2\}$	$\emptyset$	∞	P_2	3,552,439.26	Branch on y_D
3	$\{P_1, P_3, P_4\}$	$\emptyset$	∞	P_3	3,709,985.00	Branch on y_E
4	$\{P_1, P_4, P_5, P_6\}$	$\emptyset$	∞	P_6	3,920,000.00	Update BEST & BESTCOST
5	$\{P_1, P_4, P_5\}$	$\{0,0,0,1,0\}$	3,920,000.00	P_1	3,927,992.49	Continue
6	$\{P_4, P_5\}$	$\{0,0,0,1,0\}$	3,920,000.00	P_4	4,071,420.00	Continue
7	$\{P_5\}$	$\{0,0,0,1,0\}$	3,920,000.00	P_5	4,185,000	Continue
8	$\emptyset$	$\{0,0,0,1,0\}$	3,920,000.00			Terminate

P_0 = Original problem   Solution to linear programming relaxation: $\{0.00, 0.00, 0.42, 0.15, 0.43\}$

$P_1 = \{P_0, y_C = 1\}$ Solution to linear programming relaxation: $\{0.00, 0.00, 1, 0.15, 0\}$

$P_2 = \{P_0, y_C = 0\}$ Solution to linear programming relaxation: $\{0.00, 0.00, 0.00, 0.57, 0.43\}$

$P_3 = \{P_2, y_D = 1\}$ Solution to linear programming relaxation: $\{0.00, 0.00, 0.00, 1.00, 0.20\}$

$P_4 = \{P_2, y_D = 0\}$ Solution to linear programming relaxation: $\{0.00, 0.7, 0.00, 0.00, 0.93\}$

$P_5 = \{P_3, y_E = 1\}$ Solution to linear programming relaxation: $\{0.00, 0.00, 0.00, 1.00, 1.00\}$

$P_6 = \{P_3, y_E = 0\}$ Solution to linear programming relaxation: $\{0.00, 0.00, 0.00, 1.00, 0.00\}$

promising at that stage. As a result, a greedy heuristic may take suboptimal decisions and end up with a suboptimal solution.

Consider the uncapacitated facility location problem from Section 5.1. For this problem, the aim of any solution heuristic is to terminate with a set of warehouse sites at which to rent warehouses. Once that set of locations is determined, the heuristic only has to ensure that each cable laying site's requirement is met from the warehouse located closest to it. To do this, at each iteration the greedy heuristic makes a decision which looks most promising at that point in its execution.

For uncapacitated facility location problems, let the set of sites at which the greedy heuristic would suggest that we rent warehouses be S. Initially, the greedy heuristic starts with $S = \emptyset$. The transportation cost of this solution is taken to be a very high value, say M. At the beginning of the ith iteration of the greedy algorithm, we assume that $S = S_i$. During the iteration, the greedy algorithm finds out the effect of adding one more location to S_i from the locations that were not already in S_i. Let us consider a location k such that $k \notin S_i$. If k is to be included in S_i, the setup cost of the solution thus formed would increase by the rent of a warehouse at k. However, if there are some cable laying sites that are closer to k than to any other location in S_i, the total transportation cost would decrease. Therefore, the net effect of including location k in the set S_i would be beneficial if the reduction in transportation cost is greater than the increase in the setup costs. At this iteration, the greedy algorithm evaluates all locations for which the effect of adding the location is beneficial, and chooses to add the location for which the benefit is the largest. If there are no locations for which the effect of adding the location is beneficial, the greedy algorithm outputs the current set S, and terminates.

Let us illustrate the working of the greedy algorithm on the uncapacitated location problem described in Section 5.1. Let us assume that $M = 10,000,000$. At the start of the first iteration, the set S is empty. Let us now look at the effect of adding a warehouse location to this set. Table 5.9 presents the benefits of adding each of the warehouse locations to the current set S. Clearly, the benefit of adding location D is

Table 5.9. Benefits from the first iteration of the greedy heuristic

Location	A	B	C	D	E
Increase in setup cost (in €)	1,250,000	1,500,000	1,000,000	500,000	750,000
Decrease in transportation cost (in €)	6,464,000	6,020,000	6,862,000	6,580,000	6,640,000
Benefit (in €)	5,214,000	4,520,000	5,862,000	6,080,000	5,890,000

the maximum. So the greedy heuristic adds this location to S, and moves on to the second iteration.

The second iteration starts with $S = \{D\}$. The setup cost for this solution is €500,000 and the transportation cost is €3,420,000. The greedy algorithm now examines the benefit of adding one of the locations A, B, C, and E to the set S. Table 5.10 presents the benefits of adding each of the warehouse locations to the current set S. Since none of the additions is beneficial, the greedy heuristic outputs the

Table 5.10. Benefits from the second iteration of the greedy heuristic

Location	A	B	C	E
Increase in setup cost (in €)	1,250,000	1,500,000	1,000,000	750,000
Decrease in transportation cost (in €)	239,000	296,000	567,000	485,000
Benefit (in €)	-1011000	-1,204,000	-433,000	-265,000

current set $S = \{D\}$, and terminates.

In this example, the greedy heuristic outputs the optimal solution to the uncapacitated facility location problem, i.e., the best alternative is to open a warehouse at location D, and to supply all demands from it. However, the greedy heuristic is not guaranteed to always output optimal solutions.

5.5 Other Facility Location Problems

5.5.1 The competitive facility location problem

Consider two decision makers, each of whom have to locate stores in a locality. Both the decision makers want to maximize the number of people coming to their stores. Assume that people go to the store which is closest to them. In this case, a decision maker who opens the first store is at a disadvantage; she does not know where the second decision maker would open his store, while the second decision maker would be able to open his store with the knowledge of where the first decision maker has opened her store. In such cases, the first decision maker would consider various options of locations for her store. For each of these locations, she will find out the maximum share of the market that the second decision maker would have given the location of the first store. She would then choose that location for which the second decision maker's maximum market share would be the smallest possible. This problem and its extensions are known as *competitive location problems.*[1]

5.5.2 The multi-objective facility location problem

Consider a situation in which we need to locate a facility, like a bus terminus, inside a city. Due to the noise generated by such a terminus, residents would like the terminus to be located far away from their homes. However, since they have to use the bus services, they would not like the terminus to be located too far away. Hence, the city planners have to consider a location problem with two conflicting objectives instead of the problems considered earlier in this chapter, all of which have a single

---

[1] For more details on this problem, see H.A. Eiselt, G. Laporte, and J.-F. Thisse, Competitive location models: A framework and bibliography. Transportation Science 27, (1993), pp.44–54.

objective. Such facility location problems with multiple conflicting objectives are called *multi-objective facility location problems*.[2]

5.6 Exercises on Facility Location Problems

Companies like GTC require to solve network location problems in a lot of situations. In addition to locating service stations for their technicians to maintain a certain service level, and to decide on pick-up points for their employee bus service, companies now use such problems to model situations for optimizing their global supply chains.

Problem 5.1. Locating Service Stations
The Services Department of GTC manages the relationships between the company and its customers. The Gold Service, a 24-hour service for private customers, is described in Chapter 1. However, in addition to private customers, companies also use the services of GTC for reliable telecommunication systems. To cater to its corporate demand, GTC is planning to launch the Titanium Service project, whereby, whenever a company with a Titanium Service contract reports a problem, a technician will make a site call within 60 minutes.

Titanium Service will be implemented in the region schematically depicted in Figure 5.10. The nodes in this figure represent the cities. There are 20 cities, and in each city there are one or more companies located. An edge between two cities denotes a highway connection between the two, and the number next to the connection is the estimated travel time.

The companies in the region that are highly interested in the Titanium Service project are located in the cities 2, 3, 4, 5, 7, 8, 10, 11, 13, 14, 15, and 18. In order to be able to make the site calls to these twelve cities within one hour, GTC is planning to open a number of technician facilities in the region. These facilities should be located in the cities. GTC wants to know the number of technician facilities needed, and the location(s) of these facilities.

(a) Determine the minimum number of technician facilities needed for servicing the cities 2, 3, 4, 5, 7, 8, 10, 11, 13, 14, 15, and 18, and the location(s) of these facilities for the region of Figure 5.10. What is the maximum response time, and which of the twelve cities concerns this?

(b) The response time of one hour is only a suggestion. How does the answer to part (a) change if the response time is set to 30, 45, 75, or 90 minutes, respectively?

(c) The highway connection between the cities 5 and 10 is one of the main routes in the region. As a result, the travel time on this segment is most of the time two times the estimate given in Figure 5.10. What is the consequence for the maximum response time of the solution to part (a) when the travel time between the cities 5 and 10 is twice as much as shown in Figure 5.10? What are now the best locations for the facilities?

---

[2] For more details on this problem, see D.M. McAllister, Equity and efficiency in public facility location. Geographical Analysis 8, (1976), pp.47–63.

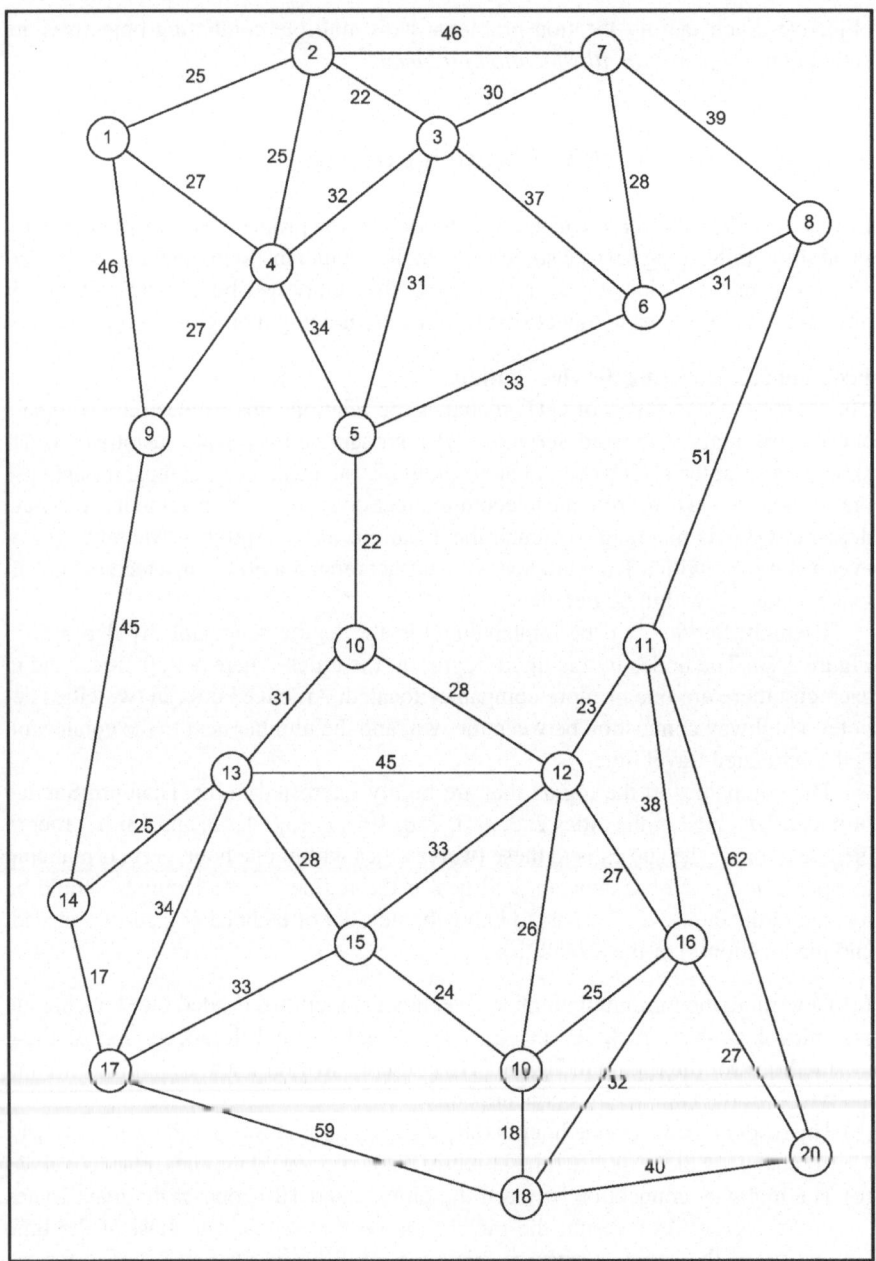

Fig. 5.10. Schematic map with 20 locations (travel times in minutes)

(d) After a second round of information gathering, it turns out that a company in city 20 wants to make use of the service as well. Does the solution to part (c) still satisfy the response time of one hour? What are now the best locations for the facilities?

(e) Some time after the implementation (the facilities are located as found in part (d)) of the project, roads 16 – 19 and 16 – 20 are temporarily closed for maintenance. GTC wants to know whether it is necessary to open one or more temporary facilities to maintain service levels.

Problem 5.2. Warehouse Location

In a certain country GTC experiences very high transportation costs. This problem is especially acute for the spool factory, where GTC produces cable spools, because it supplies all 62 demand locations in the country. A way to lower the transportation costs is by making use of warehouses. GTC wants to know where these warehouses should be located.

Figure 5.11 gives a schematic representation of the country. The nodes denote the demand locations. The number next to a node is the yearly demand for spools in units of 100 spools at the corresponding location. The edges represent the highway road system in the country. The numbers next to the edges refer to the lengths in kilometers of the corresponding road segments.

The warehouses are to be built at demand locations. The amortized cost of building a warehouse at a demand location is €2,500,000 per year. Transportation of 100 spools is €50 per kilometer. In the computations, GTC does not take the cost of demolishing a warehouse into account.

(a) GTC wants to know the minimum transportation plus warehousing costs during the next five years. Through inspection only, what is the best solution to this warehouse problem that you can find?

(b) Several years ago the government of the country implemented a large road network improvement project. As a result, within one year the following road segments will be finished (between brackets the lengths in kilometers): 5 – 49 (9.2), 19 – 39 (17.4), 19 – 40 (16.4), and 10 – 53 (5.2). The segments 46 – 47 (21.1), 18 – 30 (8.2), 11 – 14 (15.2), 14 – 22 (18.4), 22 – 58 (10.2), and 41 – 58 (10.0) will be finished within two years. Given this information, answer part (a) again.

(c) GTC wants to have a warehouse within 75 kilometers of each demand location. Does the solution to part (b) satisfy this requirement? If not, find a satisfactory solution that satisfies this requirement.

Problem 5.3. Locating Bus Stops

Each year GTC organizes a bus tour of all important GTC facilities in the region for all employees hired that year. Figure 5.12 shows a schematic representation of the region. The nodes represent the cities in the region, while the number next to a node

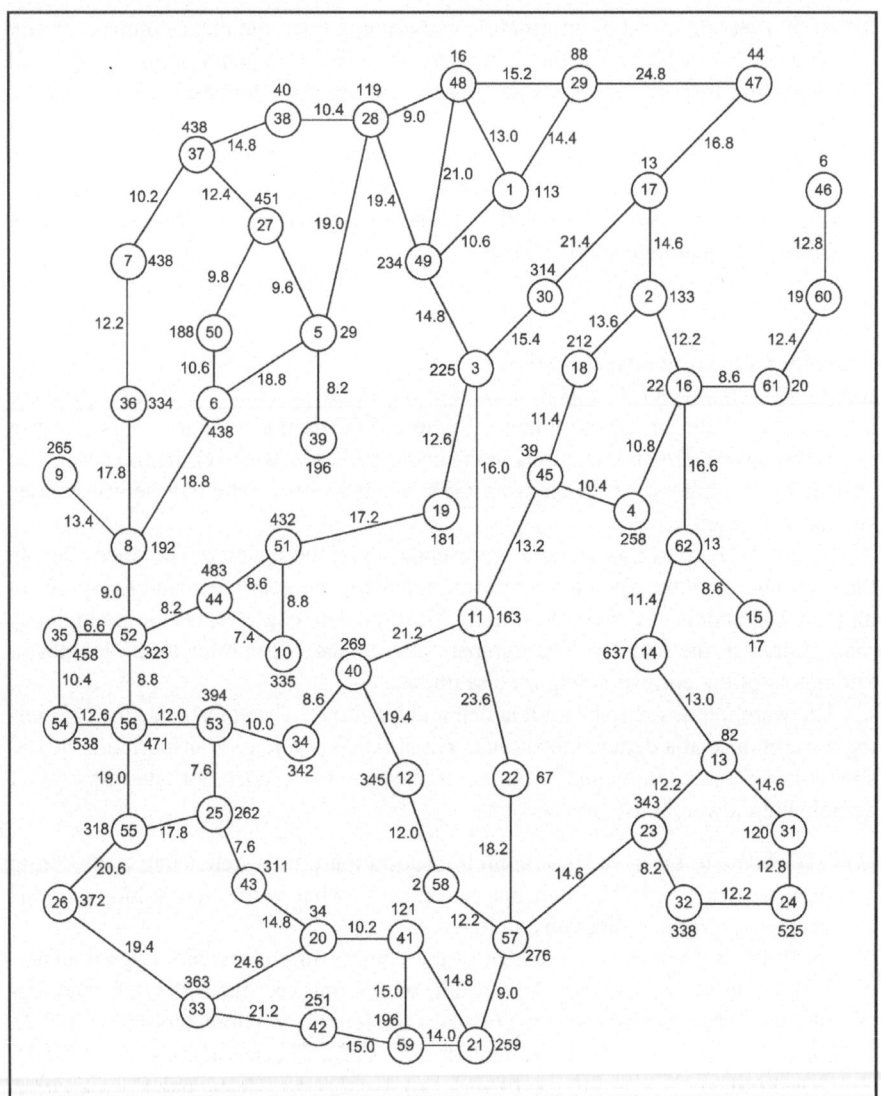

Fig. 5.11. Demands (in units of 100) and distances (in kilometers) on a road map with 62 locations

refers to the number of employees in that city that will participate in the bus tour. The lines denote the roads. The number next to a line refers to the length of that road segment (in kilometers). The facilities to be visited are located in 12, 15, 16, and 31. The tour will be a round trip, i.e., it will finish at the facility from where it started. The bus company is located at city 29.

On the morning of the day of the tour, the new employees travel to the nearest gathering point. A GTC bus will go to this point as well. When everyone is seated in the bus, the bus goes to the nearest facility to be visited and starts the tour. After the tour is finished, the bus goes back to the gathering point. From there the employees go home and the bus goes back to city 29.

The employees are compensated for traveling between their home towns and gathering point at the rate of €0.65 per kilometer. GTC has to pay the bus company €1.00 per kilometer. The total cost for the tour for GTC is the sum of the compensations paid to the employees and the cost of the bus. The question for the company therefore is, in which city to locate the gathering point such that total costs are minimized?

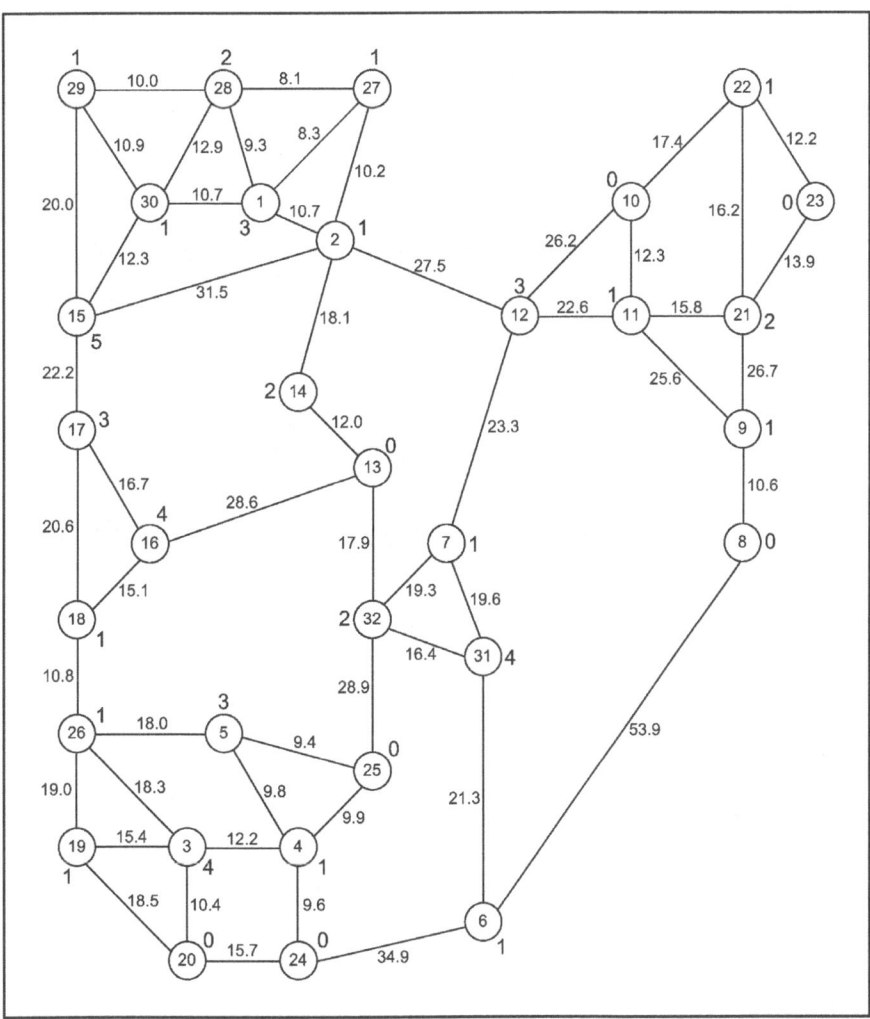

Fig. 5.12. New employees and distances (in kilometers) on a road map with 32 locations

(a) Formulate the choice of the gathering point as a 1-median problem.
(b) What is the best location for the gathering point? What are the total costs involved?
(c) Due to maintenance activities, the road segments $18 - 26$ and $25 - 32$ will be closed for a certain period. Does this alter the choice of the gathering point? What are the total costs now?
(d) GTC is wondering whether it would be more cost effective to use two buses. Can the choice of two gathering points be formulated as a 2-median problem? Explain the answer.

An alternative plan is that the bus will pick up people at multiple locations. The bus goes via a specified route from city 29 to one of the four facilities to be visited. The bus will halt at all cities in between. The employees are free to choose where they want to board the bus.

(e) Suppose that the bus follows the route $29 - 30 - 1 - 2 - 12$. This means that the employees can go to either city 29, 30, 1, 2, or 12. What are the total costs for this route?
(f) Find by inspection, a route for the bus, such that the total costs are within €1100.

Problem 5.4. Switching Point Location
In Problem 5.1, the question was to design a cable network. The network needs a number of switching points for assuring a certain degree of reliability. In this problem we examine the location of switching points for the network found in Problem 5.1(a).

(a) Assuming that a switching point can only be located at one of the 50 locations, what is the best location if GTC considers locating only one switching point? What is the distance from the switching point to the location farthest from it? Call this distance D_1. If GTC considers locating two switching points where would they be located? What would be the maximum distance between any location and any switching point? Call this distance D_2.
(b) Assume now that the switching points can be located anywhere on the cable network. Answer the same questions as in part (a). Call the distances D_1^* and D_2^*, respectively.
(c) Check from your answers of parts (a) and (b) that $D_1 \geq D_1^* \geq D_2$. Prove that this relation holds for all networks with at least two nodes.
(d) Is it true for the network found in Problem 5.1(a) that $D_2 \geq D_2^* \geq D_3$? Consider a general network with n nodes. Prove or disprove that $D_i \geq D_i^* \geq D_{i+1}$ for all i with $i = 1, \ldots, n-1$.

6

Cyclic Routing on Networks

6.1 Introduction

Every year, a team of executives in the corporate office of GTC visits each of the five regional offices to inspect the work at those offices. This trip is a round trip in which the officers visit each of the region offices exactly once before returning to the corporate office. The regional offices are located in cities connected to each other by air. The cost of flying between each pair of cities is given in Table 6.1. C denotes the city in which the corporate office is located, and R1 through R5 denote the cities in which the regional offices are located.

Table 6.1. Air fares between cities (in €)

	C	R1	R2	R3	R4	R5
C	—	500	650	475	525	925
R1		—	625	750	550	825
R2			—	750	575	425
R3				—	450	725
R4					—	950
R5						—

The total cost of air travel would obviously depend on the route that the team of executives takes. For instance, if they travel from C to R1 to R2 to R3 to R4 to R5 and back to C, then the cost per executive would be €4200. On the other hand, if they travel from C to R3 to R5 to R2 to R4 to R1 and back to C, then the cost per executive reduces to €3250. The financial department at GTC wants to find out how to route the travel of executives so that the total travel costs are minimized.

The problem of routing a round trip through all cities in a network visiting each city exactly once before returning to the city of origin is called the *traveling salesman problem*. This is one of the most well-known network problems. A solution to the problem, i.e., a round trip through all cities is called a *tour*. The cost of a tour is the

G. Sierksma and D. Ghosh, *Networks in Action: Text and Computer Exercises in Network Optimization*, International Series in Operations Research & Management Science 140, DOI 10.1007/978-1-4419-5513-5_9, © Springer Science + Business Media, LLC 2010

sum of the costs of all links on the tour. So the traveling salesman problem is one of finding the least cost tour on a network. If the intercity distances are such that the cost of going from city i to city j is the same as the cost of going from city j to city i for all pairs of cities i and j, then the problem is called a *symmetric traveling salesman problem*. If this is not the case, then the problem is called an *asymmetric traveling salesman problem*. A closely associated problem is called the *Hamiltonian cycle problem*. In this problem, we are given a network and asked whether or not a *Hamiltonian cycle* i.e., a tour in the network visiting each city exactly once, exists in that network.

Now consider a different problem. The research group of GTC in city C has developed a new prototype which it wants to distribute to the research groups in the regional offices. They want to load five prototypes and send them to the offices in a truck. The truck must therefore visit each regional office exactly once. It does not have to return to the corporate office at the end of its journey. GTC knows the distance between the cities on the road network. These distances are given in Table 6.2.

Table 6.2. Distances between cities along the road network (in kilometers)

	C	R1	R2	R3	R4	R5
C	—	437	516	356	439	718
R1	420	—	519	599	473	640
R2	538	535	—	622	430	364
R3	353	613	633	—	326	571
R4	429	442	469	366	—	734
R5	754	699	341	563	723	—

If, for example, GTC decides that the truck should go from city C to R1 to R2 to R3 to R4 to R5, then the total distance is 2638 kilometers, while if they decide that the truck should go from C to R3 to R5 to R2 to R4 to R1, then the distance covered, namely 2498 kilometers, is less. The financial department at GTC would be interested in computing a shortest route through all the cities.

This problem of finding a shortest route through all the cities in a network, covering each city exactly once, is known as a *weighted Hamiltonian path problem*. It is a weighted version of the *Hamiltonian path problem* in which we are given a network and are required to find out whether or not there exists a *Hamiltonian path* in the network, i.e., a path in the network visiting each city exactly once. The difference between the Hamiltonian path problem and the traveling salesman problem is only that in the weighted Hamiltonian path problem, we are not required to return to the city of origin to end the route. It is also different from the shortest path problem, since in the Hamiltonian path problem, we are required to visit *every* city in the network, and that too, exactly once.

Notice that a weighted Hamiltonian path problem can easily be converted into a traveling salesman problem. To do this we need to add a dummy node to the network on which the weighted Hamiltonian path problem is defined, and connect it to all the nodes in the network through edges that have zero lengths. It is easy to see that all Hamiltonian paths correspond to Hamiltonian cycles in the augmented network. Also, since the edges joining the dummy node to the other nodes in the Hamiltonian cycle have zero length, the cost of the Hamiltonian cycle in the augmented network is the same as the Hamiltonian path that is formed by removing the dummy node and the two edges incident on it from the cycle. Therefore, if we find an optimal solution to the traveling salesman problem in the augmented network, and remove the dummy node and the two edges incident on it from the solution, then we have an optimal solution to the weighted Hamiltonian path problem.

We next consider a third routing problem. The distribution system of GTC is a two-tier system. GTC delivers supplies from the regional office to a central depot in an area, and then the staff at the central depot delivers them to stores in the area. The location of the depot and nine stores in a particular area are shown schematically in Figure 6.1. The depot is labeled D, and the nine stores are labeled 1 through 9.

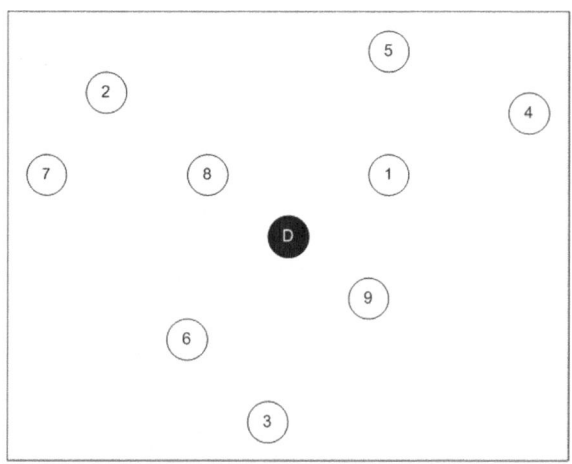

Fig. 6.1. The regional distribution problem

The depot has three trucks at its disposal to distribute the supplies to the stores. The depot staff would like to route the trucks in such a way as to minimize the total distance traveled by these trucks. Since the cost of traveling is directly proportional to the distance traveled, their objective would also be achieved by minimizing the distribution cost. We would like each of the stores to be visited exactly once during the distribution process.

This problem is known as the *vehicle routing problem*. A solution to the vehicle routing problem is a routing schedule for each truck. Figure 6.2 illustrates a possible solution to the problem. According to this solution, one truck starts at the depot, supplies stores 1, 4, and 5, in that order and returns to the depot. A second truck supplies stores 8, 2, and 7, in that order and returns to the depot, while the third truck supplies stores 9, 3, and 6 in that order and returns to the depot. Notice that if

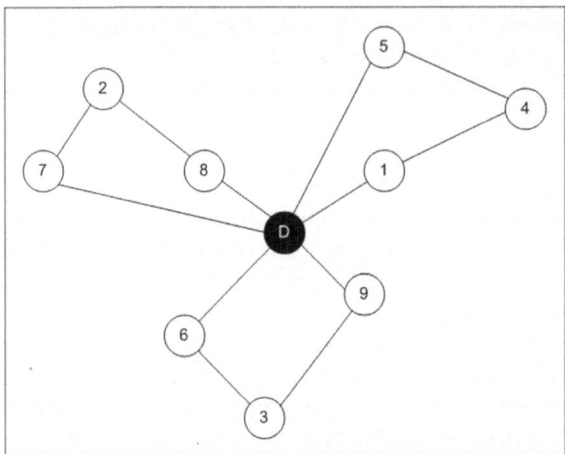

Fig. 6.2. A solution to the regional distribution problem

the depot had only one truck, then the solution to the routing problem would be the solution to a traveling salesman problem defined on D and the nine store locations.

Even in case there are more than one trucks at the depot's disposal, the problem can be transformed into a traveling salesman problem. Since there are three trucks, we make five copies of the node in the network corresponding to the depot. (If there are n vehicle at the depot, then we make $2n - 1$ copies of the depot in the network.) The depot and its five copies can be thought of as six nodes of the network, connected to each other by edges of zero length. These six nodes are represented in Figure 6.3 as the six nodes within the circle with broken lines. Each of the six copies are connected to the rest of the stores exactly in the same way as the depot originally was. A traveling salesman problem defined on this new network would yield a solution to the vehicle routing problem. Figure 6.3 for example illustrates the solution in Figure 6.2 on this new network. The way to obtain the individual routings from the solution to the traveling salesman problem is to recombine the six nodes corresponding to the depot into a single node, and to mark off the route between two consecutive visits to the depot as the route for a truck.

The problems described above all belong to the class of *node routing problems*. Any solution to these problems requires a routing through each node of the network on which the problem is defined. There are also routing problems that need to involve every edge in an undirected network, and every arc in a directed one. These problems are called *arc routing problem*. Consider, for example, the network given in Figure 6.4. In such networks, the problem of finding a routing that traverses every edge of the network at least once, comes back to the node at which the route started, and has the minimum number of edges possible is called the *Chinese postman problem*. It is a classical example of an arc routing problem. An example of a solution to the Chinese postman problem on the network in Figure 6.4 is to go from A to B to D to C to B to D and then return to A. Notice that we have traversed edge B – D twice

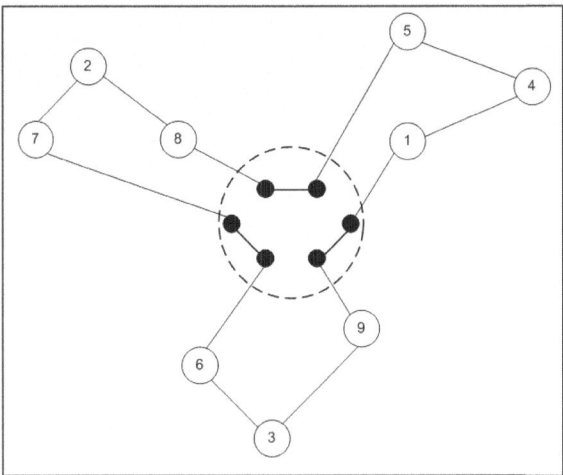

Fig. 6.3. Transforming a VRP instance into a TSP instance

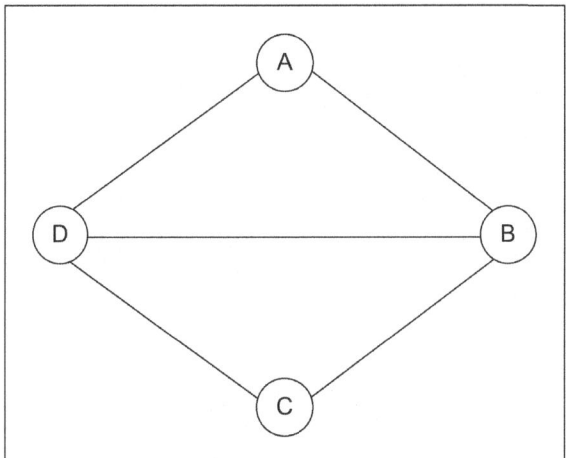

Fig. 6.4. A network to illustrate the Chinese postman problem

in our route. Therefore, this route would be inferior to a route that traverses each edge exactly once. Unfortunately, not all networks admit solutions to Chinese postman problems in which we traverse each edge exactly once. Networks that do admit such solutions must possess the *Eulerian property*, which means that each node in the network has an even degree. Since the network in Figure 6.4 does not possess this property, since nodes B and D have odd degree, we can safely say that there cannot be a Chinese postman tour that traverses each edge in this network exactly once. It can be shown that in any network the maximum number of edges traversed in an optimal solution to the Chinese postman problem is twice the number of edges in the network. The network in Figure 6.5 is an example of a network for which the

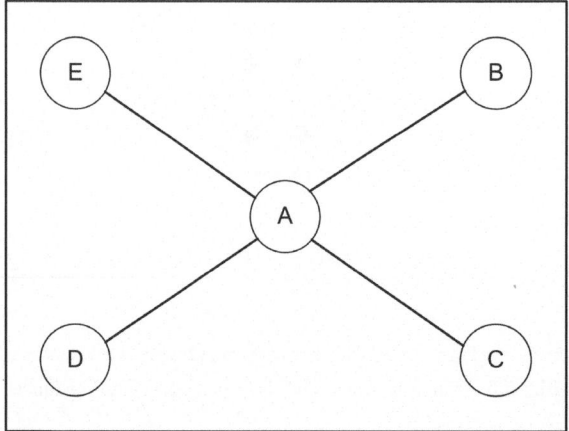

Fig. 6.5. A network that is bad for the Chinese postman problem

optimal solution to the Chinese postman problem needs to traverse each edge exactly twice.

If the Chinese postman problem is defined on a weighted graph, with the weight on each edge signifying the length of the edge, then it is called a *weighted Chinese postman problem*. The objective in a weighted Chinese postman problem is to find a tour of minimum length that traverses each edge exactly once.

6.2 Applications

The traveling salesman problem forms the basis of many real world applications, some of which are described in this section. The vehicle routing problem described in the previous section is an application in itself, although real world implementations of vehicle routing problems have to deal with many more practical constraints. Similarly, arc routing problems have direct application in public systems. The following are some applications of routing problems.

6.2.1 Manufacturing of printed circuit boards

Printed circuit boards (PCBs) have become an electronic part seen very commonly in most electronic equipment. It consists of conductors that are "printed" on a board. The board also has holes for attaching the pins of integrated circuit chips. The holes could have different diameters. While manufacturing PCBs, holes of the same diameter are drilled consecutively in batches. In order to produce PCBs at a fast rate, one needs to optimize the sequence in which the holes of a batch have to be drilled on a PCB. To do this, the drill bit needs to be attached to the drilling machine at the tool box, then the holes need to be drilled, and the drill bit needs to be released at the tool box. The sequence in which the holes have to be drilled can be computed by modeling the problem as a traveling salesman problem. A complete graph is constructed with the tool box and the holes that have to be drilled as the nodes of the graph. Each pair of nodes is then connected with an arc whose cost is proportional to the time required to move between the locations corresponding to the end nodes. A minimum cost tour in this graph is the most economical way of drilling a batch of holes on the PCB.

6.2.2 Order picking in warehouses

Large warehouses stock large numbers of items divided into sections. Each section houses similar items. Most orders from warehouses include items from different sections. For a warehouse to fill orders efficiently, a manager needs to compute the sequence in which to pick items from the different sections so that an order can be filled in minimum time. This problem can be modeled as a traveling salesman problem. The problem is defined on a complete graph, in which one of the nodes corresponds to the entry/exit point of the warehouse, and the other nodes correspond to the sections which contain items required to fill the order. The cost of each arc in the graph corresponds to the time required to move from the location denoted by the tail of the arc to the location denoted by the head of the arc. Starting from the node correesponding to the entry/exit of the warehouse, the sequence of nodes in an optimal tour through this graph is the sequence of sections from which items need to be picked to fill the order.

6.2.3 Postal delivery routing

Any postal delivery mechanism consists of three main tasks, sorting mails based on postcodes at a central facility, packing mail for a single postcode to a facility at that postcode, and physically delivering the mail in the postcode. Among the three activities, the last is the most time consuming. Therefore, an efficient postal delivery system tries to make the physical delivery of mail as efficient as possible. For purposes of delivery, the area corresponding to a postcode is divided into smaller areas, and each area is assigned to a single postal delivery person. The route that the person should take to deliver the mail can be modeled as a weighted Chinese postman problem. The road network in the area forms the graph for the problem, and the weights on each of the roads correspond to the time taken to traverse the road segment. The minimum weight solution to the weighted Chinese postman problem on this graph denotes the route that a postal delivery worker should take to deliver mail in the area.

6.3 (Integer) Linear Programming Formulations

All the network routing problems described in the previous section can be formulated as linear programming problems. Since all the node routing problems described can be easily reformulated as a traveling salesman problem, we will describe formulations of the traveling salesman problem only for node routing problems. We will of course describe a formulation of the Chinese postman problem as an example of formulations for arc routing problems.

6.3.1 The traveling salesman problem

There are several formulations for the traveling salesman problem. In what follows we provide two formulations for the problem. These two formulations serve to illustrate the diverse ways in which the same problem can be formulated.

In both formulations, we first convert any edges present in the network into arcs. As with other network problems, if there is an edge between cities i and j in a network with length d_{ij}, it is converted to two arcs, one from i to j and another from j to i. Both the arcs thus formed have length d_{ij}, and we have constraints that restrict us to include at most one of the two arcs in our optimal solution. This step is of course unnecessary if the network that we start out with is directed.

In the first formulation, we use decision variables y_{ij} for each arc $i \rightarrow j$ which assumes a value of 1 if this arc is included in the optimal solution, and 0 otherwise. Let is denote the network as $N = (V, A, d)$, where V denotes the set of cities, A denotes the set of interconnecting arcs between the cities, and d_{ij} denotes the length of arc $i \rightarrow j \in A$. The length of a tour is the sum $\sum_{i \rightarrow j \in A} d_{ij}$, and the objective of the formulation is to

$$\text{Minimize} \sum_{i \rightarrow j \in A} d_{ij} y_{ij}. \tag{6.1}$$

In a solution to the traveling salesman problem, each city in the network is visited exactly once. Therefore, for each city i in the network, there should be exactly one arc of the network to i and one arc of the network from i. For each $i \in V$, this leads us to the following two constraints.

$$\sum_{j: j \rightarrow i \in A} y_{ji} = 1, \text{ and} \tag{6.2}$$

$$\sum_{j: i \rightarrow j \in A} y_{ij} = 1. \tag{6.3}$$

At first glance it may seem that these two sets of constraints are both necessary and sufficient to describe tours in the traveling salesman problem. But even though these constraints are necessary, they are not sufficient to ensure that solutions to the formulation are restricted only to tours. For example, the solution illustrated in Figure 6.6 for a traveling salesman problem instance described on six cities A through F is not a tour, although it does obey the constraint sets (6.2) and (6.3). The loops in the solution illustrated in the figure are called *subtours* and our formulation needs

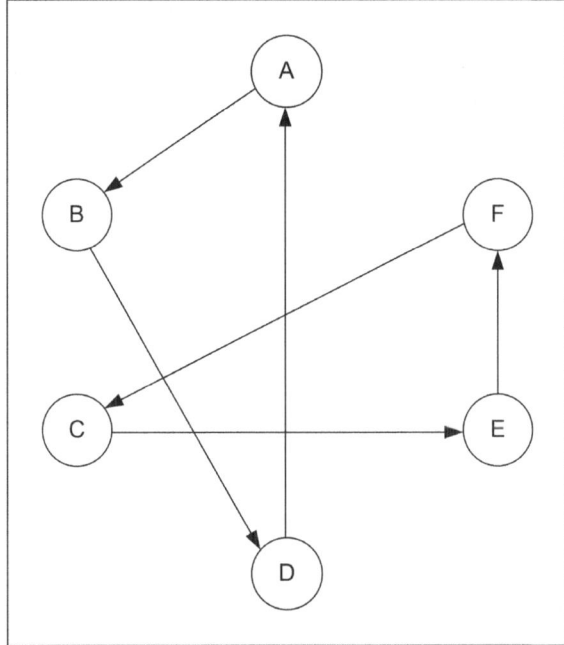

Fig. 6.6. Subtours in a traveling salesman problem

to add subtour elimination constraints to eliminate all solutions that are not tours. There are several ways of including subtour elimination constraints, most of which are reminiscent of similar constraints in the linear programming formulation of minimum spanning trees. We here illustrate two ways of including subtour elimination constraints. In the first method, we specify that for all subsets S of V, except for the empty set and the set V itself, there should be at least one arc exiting the set. In case subtours exist, a set of nodes forming one of the subtours would fail to satisfy this condition. Hence, we have the following set of subtour elimination constraints.

$$\sum_{i,j:i \rightarrow j \in A,\, i \in S,\, j \notin S} y_{ij} \geq 1 \qquad \text{for all } \emptyset \subset S \subset V. \qquad (6.4)$$

Notice however, that for all subsets S with $|S| = 1$, these constraints are subsumed by the constraint set (6.3), and for all subsets S with $|S| = |V| - 1$ they are subsumed by the constraint set (6.2). So we only need to implement these constraint for all subsets S for which $2 \leq |S| \leq |V| - 2$. The complete formulation described above is given in Figure 6.7.

Alternatively, we can stipulate that for all subsets S of V, except for the empty set and the set V itself, the number of arcs in an optimal solution between two nodes within S should be at most $|S| - 1$. In case subtours exist, a set of nodes forming one of the subtours would fail to satisfy this condition, since the subtour will ensure that there are $|S|$ arcs within the set. Hence, we have the following alternate set of subtour

Minimize

$$z = \sum_{i \to j \in A} d_{ij} y_{ij}$$

Subject to

$$\sum_{j:j \to i \in A} y_{ji} = 1 \qquad \text{for all } i \in V$$

$$\sum_{j:i \to j \in A} y_{ij} = 1 \qquad \text{for all } i \in V$$

$$\sum_{i,j:i \to j \in A,\, i \in S,\, j \notin S} y_{ij} \geq 1 \qquad \text{for all } \emptyset \subset S \subset V$$

$$y_{ij} \in \{0,1\} \text{ for each } i, j \in V$$

Fig. 6.7. First linear programming formulation of the traveling salesman problem

elimination constraints.

$$\sum_{i,j:i \to j \in A,\, i,j \in S} y_{ij} \leq |S| - 1 \qquad \text{for all } \emptyset \subset S \subset V. \qquad (6.5)$$

It is easy to see that for these subtour elimination constraints also, it is sufficient to formulate the constraints for all subsets S for which $2 \leq |S| \leq |V| - 2$. The complete formulation with the alternative method of modeling subtour elimination is given in Figure 6.8.

In order to illustrate this formulation, we will use the air travel problem from Section 6.1. The network in this problem consists of six nodes, C, R1, ..., R5. Each inter-city connection can be represented as an arc, so that the network is a directed network. The cost of an arc is the cost of the connection between the two cities that the arc connects. Notice that this is an instance of the symmetric traveling salesman problem. We use the constraint set (6.4) to eliminate subtours. The formulation is given in Figure 6.9.

An optimal solution to this problem is obtained by setting $y_{C,R3}$, $y_{R1,C}$, $y_{R2,R5}$, $y_{R3,R4}$, $y_{R4,R2}$, and $y_{R5,R1}$ to 1 and all the other variables to 0. This solution corresponds to the tour from C to R3 to R4 to R2 to R5 to R1 and then back to C. The total cost of the tour is €3,250.

We next describe a second formulation of the traveling salesman problem. In this formulation we assume that we start from a city v_0 and then visit each of the other cities in sequence until we come back to v_0 at the end of the tour. Thus a tour is composed of several "legs". In each leg we start from a city (say i) in the sequence and end at the next city (say j) in the sequence. The cost of a leg is the cost of the arc from i to j. The decision variables in this formulation therefore are of the form y_{ijk} which assumes a value of 1 if the arc from i to j makes up the kth leg of the tour,

Minimize

$$z = \sum_{i \to j \in A} d_{ij} y_{ij}$$

Subject to

$$\sum_{j:j \to i \in A} y_{ji} = 1 \qquad \text{for all } i \in V$$

$$\sum_{j:i \to j \in A} y_{ij} = 1 \qquad \text{for all } i \in V$$

$$\sum_{i,j:i \to j \in A, \, i,j \in S} y_{ij} \leq |S| - 1 \qquad \text{for all } \emptyset \subset S \subset V$$

$$y_{ij} \in \{0,1\} \text{ for each } i,j \in V$$

Fig. 6.8. Second linear programming formulation of the traveling salesman problem

and a value of 0 otherwise. The objective in the formulation is thus

$$\text{Minimize} \sum_{k=1}^{|V|} \sum_{i,j:i \to j \in A} d_{ij} y_{ijk}. \tag{6.6}$$

In this formulation too, we have the standard constraints that ensure that we enter and leave a city exactly once. For each city $i \in V$ these constraints are of the form

$$\sum_{k=1}^{|V|} \sum_{j:j \to i \in A} y_{jik} = 1, \text{ and} \tag{6.7}$$

$$\sum_{k=1}^{|V|} \sum_{j:i \to j \in A} y_{ijk} = 1. \tag{6.8}$$

We also specify that there are exactly $|V|$ legs in the tour, using the constraint

$$\sum_{k=1}^{|V|} \sum_{i,j:i \to j \in A} y_{ijk} = |V|, \tag{6.9}$$

and that at each leg of the tour only one arc can be traversed, with the set of constraints

$$\sum_{i,j:i \to j \in A} y_{ijk} = 1 \text{ for each } k = 1, \dots, |V|, \tag{6.10}$$

and also that each arc can be traversed in at most one leg of the tour, with the following set of constraints:

Minimize

$$z = 500y_{C,R1} + 650y_{C,R2} + \cdots + 525y_{C,R4} + 925y_{C,R5} +$$
$$500y_{R1,C} + 625y_{R1,R2} + \cdots + 550y_{R1,R4} + 825y_{R1,R5} + \cdots +$$
$$925y_{R5,C} + 825y_{R5,R1} + \cdots + 725y_{R5,R3} + 950y_{R5,R4}$$

Subject to

$$y_{R1,C} + y_{R2,C} + \cdots + y_{R4,C} + y_{R5,C} = 1 \quad \text{(Constraint (6.2) for C)}$$
$$y_{C,R1} + y_{R2,R1} + \cdots + y_{R4,R1} + y_{R5,R1} = 1 \quad \text{(Constraint (6.2) for R1)}$$

There are four more similar constraints for R2, R3, R4, and R5.

$$y_{C,R1} + y_{C,R2} + \cdots + y_{C,R4} + y_{C,R5} = 1 \quad \text{(Constraint (6.3) for C)}$$
$$y_{R1,C} + y_{R1,R2} + \cdots + y_{R1,R4} + y_{R1,R5} = 1 \quad \text{(Constraint (6.3) for R1)}$$

There are four more similar constraints for R2, R3, R4, and R5.

$$y_{C,R2} + y_{C,R3} + y_{C,R4} + y_{C,R5} + y_{R1,R2} + \qquad \text{(Constraint (6.4)}$$
$$y_{R1,R3} + y_{R1,R4} + y_{R1,R5} \geq 1 \quad \text{for } S = \{C, R1\})$$
$$y_{C,R1} + y_{C,R3} + y_{C,R4} + y_{C,R5} + y_{R2,R1} + \qquad \text{(Constraint (6.4)}$$
$$y_{R2,R3} + y_{R2,R4} + y_{R2,R5} \geq 1 \quad \text{for } S = \{C, R2\})$$

There are thirteen more similar constraints for different S, $|S| = 2$.

$$y_{C,R3} + y_{C,R4} + y_{C,R5} + y_{R1,R3} + y_{R1,R4} + \qquad \text{(Constraint (6.4)}$$
$$y_{R1,R5} + y_{R2,R3} + y_{R2,R4} + y_{R2,R5} \geq 1 \quad \text{for } S = \{C, R1, R2\})$$
$$y_{C,R1} + y_{C,R4} + y_{C,R5} + y_{R2,R1} + y_{R2,R4} + \qquad \text{(Constraint (6.4)}$$
$$y_{R2,R5} + y_{R3,R1} + y_{R3,R4} + y_{R3,R5} \geq 1 \quad \text{for } S = \{C, R1, R3\})$$

There are eighteen more similar constraints for different S, $|S| = 3$.

$$y_{C,R4} + y_{C,R5} + y_{R1,R4} + y_{R1,R5} + \qquad \text{(Constraint (6.4)}$$
$$y_{R2,R4} + y_{R2,R5} + y_{R3,R4} + y_{R3,R5} \geq 1 \quad \text{for } S = \{C,R1,R2,R3\})$$
$$y_{C,R2} + y_{C,R5} + y_{R1,R2} + y_{R1,R5} + \qquad \text{(Constraint (6.4)}$$
$$y_{R3,R2} + y_{R3,R5} + y_{R4,R2} + y_{R4,R5} \geq 1 \quad \text{for } S = \{C,R1,R3,R4\})$$

There are thirteen more similar constraints for different S, $|S| = 4$.

$$y_{C,R1}, y_{C,R2}, \ldots, y_{R5,R3}, y_{R5,R4} \in \{0,1\} \quad \text{(Binary variables)}$$

Fig. 6.9. Formulation of the traveling salesman problem from Section 5.1

$$\sum_{k=1}^{|V|} y_{ijk} \leq 1 \text{ for each } i \text{ and } j \text{ such that } i \rightarrow j \in A. \qquad (6.11)$$

Subtour elimination constraints are included by stipulating that the first leg of the tour starts at a node labeled v_0, using the constraint

$$\sum_{j:v_0 \to j \in A} y_{v_0 j 1} = 1, \tag{6.12}$$

and that the last leg of the tour ends at the same node, using the constraint

$$\sum_{i:i \to v_0 \in A} y_{i v_0 |V|} = 1, \tag{6.13}$$

and that the $(k+1)$th leg of the tour starts at the same node where the kth leg ended. This last condition is implemented using the following sets of constraints.

$$\sum_{i:i \to j \in A} y_{ijk} - \sum_{l:j \to l \in A} y_{jlk+1} = 0 \text{ for each } j \in V, \text{and each } k = 2, \ldots, |V| - 1. \tag{6.14}$$

The complete formulation described above is given in Figure 6.10. It is left to the reader to implement this formulation for the problem from Section 6.1, and to compare it with the earlier formulation.

6.3.2 The Chinese postman problem

As we had mentioned in Section 6.1, an optimal solution to the Chinese postman problem traverses each edge in a network exactly once, if and only if the network is Eulerian, i.e., if and only if each node in the network has an even degree. Unfortunately, since not all networks are Eulerian, an optimal solution to the Chinese postman problem on a network may need to traverse an edge in the network more than once.

Linear programming formulations for Chinese postman problems on networks attempt to determine the minimum number of times that an optimal solution to the Chinese postman problem needs to traverse each edge in the network. To do so, it simulates multiple traversals of an edge by adding multiple copies of the edge in the network. The formulation then tries to minimize the number of copies of edges that needs to be added to the network to make it Eulerian.

In our description of the formulation, we consider a network $N = (V, E)$. The vertices in V are partitioned into two sets, V_o consisting of vertices that have odd degree in the network, and V_e consisting of vertices that have even degree. We also use a node-edge incidence matrix A of dimension $|V| \times |E|$, in which $a_{ie} = 1$ if edge e is incident on node i, and 0 otherwise. Our decision variables are integer variables y_e that denote the number of *copies* of the edge e we need to add to the network in order to make it Eulerian.

The objective is to minimize the number of edges added, i.e., to

$$\text{Minimize} \sum_{e \in E} y_e. \tag{6.15}$$

Each time a copy of an edge is added to the network, the degrees of the nodes it is incident on are increased by 1. So for the final network to be Eulerian, we need to

Minimize

$$z = \sum_{k=1}^{|V|} \sum_{i,j:i \to j \in A} d_{ij} y_{ijk}$$

Subject to

$$\sum_{k=1}^{|V|} \sum_{j:j \to i \in A} y_{jik} = 1 \qquad \text{for each } i \in V$$

$$\sum_{k=1}^{|V|} \sum_{j:i \to j \in A} y_{ijk} = 1 \qquad \text{for each } i \in V$$

$$\sum_{k=1}^{|V|} \sum_{i,j:i \to j \in A} y_{ijk} = |V|$$

$$\sum_{i,j:i \to j \in A} y_{ijk} = 1 \qquad \text{for each } k = 1, \ldots, |V|$$

$$\sum_{k=1}^{|V|} y_{ijk} \le 1 \qquad \text{for each } i,j \text{ such that } i \to j \in A$$

$$\sum_{j:v_0 \to j \in A} y_{v_0 j 1} = 1$$

$$\sum_{i:i \to v_0 \in A} y_{i v_0 |V|} = 1$$

$$\sum_{i:i \to j \in A} y_{ijk} - \sum_{l:j \to l \in A} y_{jlk+1} = 0 \qquad \text{for each } j \in V; k = 2, \ldots, |V| - 1$$

$$y_{ijk} \in \{0,1\} \quad \text{for each } i,j \text{ such that } i \to j \in A; k = 1, \ldots, |V|$$

Fig. 6.10. Second linear programming formulation of the traveling salesman problem

ensure that the total number of edges (including copies) that are incident to a node $i \in V_e$ is even, i.e., can be represented in the form $2w_i$, where w_i is an integer. This is done using the following constraint for each $i \in V_e$.

$$\sum_{e \in E} a_{ie} y_e = 2w_i. \tag{6.16}$$

Similarly, we need to ensure that for each node $j \in V_o$, the number of copies of edges that are added and are incident on j must be odd, i.e., can be represented in the form $2w_j + 1$, where w_j is an integer. This is done using the following constraint for each $i \in V_o$.

$$\sum_{e \in E} a_{je} y_e = 2w_j + 1. \tag{6.17}$$

These two sets of constraints are enough to decide how many copies of each edge need to be added to the network to make it Eulerian.

If we need to solve a weighted Chinese postman problem, the constraint sets remain the same as that in the unweighted problem. The only change that occurs in the formulation is that our objective changes to

$$\text{Minimize} \sum_{e \in E} l_e y_e \tag{6.18}$$

where l_e is the length of the edge e.

The complete formulation of the Chinese postman problem is given in Figure 6.11.

Minimize

$$z = \sum_{e \in E} y_e$$

Subject to

$$\sum_{e \in E} a_{ie} y_e = 2w_i \qquad \text{for all } i \in V_e$$

$$\sum_{e \in E} a_{je} y_e = 2w_j + 1 \qquad \text{for all } j \in V_o$$

$$y_e \in \{0, 1, 2, \ldots\} \qquad \text{for each } e \in E$$

Fig. 6.11. Linear programming formulation of the Chinese postman problem

We now illustrate this formulation on the network shown in Figure 6.4. The binary decision variables that we define are y_{AB}, y_{BC}, y_{CD}, y_{AD}, and y_{BD}, which determine how many copies of edges A – B, B – C, C – D, A – D, and B – D, respectively, we need to add to the network to make it Eulerian. We also introduce integer variables w_A, w_B, w_C, and w_D for the four nodes A, B, C, and D respectively. Set $V_o = \{B, D\}$ and $V_e = \{A, C\}$. The formulation is described in Figure 6.12.

6.4 Algorithms for Routing Problems

Here too, we separate the discussion into two parts, the first part dealing with node routing problems, and the second part dealing with arc routing problems. In node routing, we will describe algorithms for the traveling salesman problem, and in arc routing we will describe an algorithm for the Chinese postman problem.

6.4.1 Traveling salesman problem

The traveling salesman problem is a NP-hard problem. This means that we do not have an algorithm for finding an optimal solution to all instances of traveling salesman problems within reasonable time. In this regard, the traveling salesman problem

Minimize

$$z = y_{AB} + y_{BC} + y_{CD} + y_{AD} + y_{BD}$$

Subject to

$y_{AB} + y_{AD} - 2w_A$	$= 0$	(Constraint (6.16 at A)
$y_{AB} + y_{BC} + y_{BD} - 2w_A$	$= 1$	(Constraint (6.17 at B)
$y_{BC} + y_{CD} - 2w_C$	$= 0$	(Constraint (6.16 at C)
$y_{AD} + y_{BD} + y_{CD} - 2w_D$	$= 1$	(Constraint (6.17 at D)

$$y_{AB}, y_{BC}, y_{CD}, y_{AD}, y_{BD} \in \{0, 1, 2, \ldots\} \quad \text{(Integer variables)}$$

Fig. 6.12. Formulation of the Chinese postman problem example

is of a more similar level of difficulty as location problems rather than of the network flow problems or the minimum spanning tree problems. However, it would be incorrect to assume that all instances of the traveling salesman problem are difficult to solve. For example, Problem A in Figure 6.6 is much easier to solve than Problem B, although both the problems have the same number of cities. (Try to find a shortest round trip for both instances.)

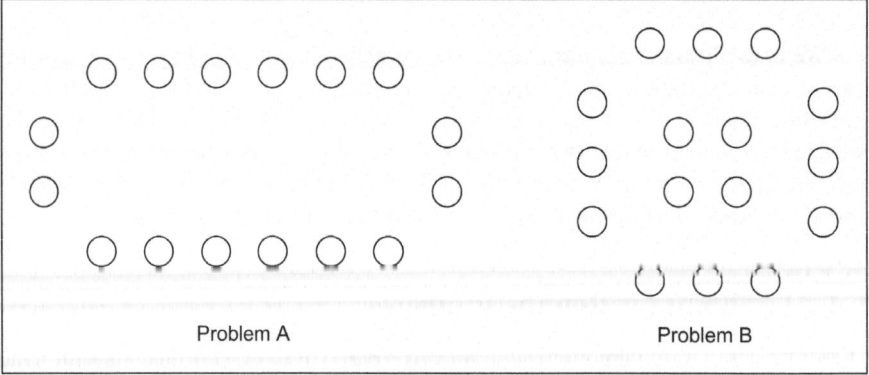

Fig. 6.13. Same sized TSP instances may vary in difficulty

Algorithms to generate an optimal tour for the traveling salesman problem typically depend on branch and bound. The basic concepts of branch and bound have been explained in detail in Chapter 4.

For the traveling salesman problem, a bound can be obtained by ignoring the subtour elimination constraints in the first formulation of the traveling salesman problem described in Section 6.3. This is simply an assignment problem as defined in Chapter 4, and can be solved efficiently using either the linear programming technique, or using more specialized algorithms.

The solution to the assignment problem that yields the bound normally contains subtours. When we choose a subproblem for the branching procedure, we look at the solution to the assignment problem defined on that subproblem. If the solution to the assignment problem contains a single tour, then that subproblem need not be considered for branching. If the subproblem contains two or more subtours, then we choose the subtour containing the smallest number of arcs. Assume that this subtour consists of m arcs. The branching procedure then creates m subproblems from this subproblem. In subproblem i, the branching procedure stipulates that the ith arc of the subtour should not be included.

As an example, let us describe the branch and bound procedure for the traveling salesman problem described in Section 6.1. We use the structures described for the branch and bound algorithm in Chapter 4. Initially LIST contains the original problem (denoted as P_0), BEST = $\emptyset$, and BESTCOST = ∞. In the first iteration, we remove P_0 from LIST, and solve an assignment problem on this network. The solution to the assignment problem on this network contains the three subtours C→R1→C, R2→R5→R2, and R3→R4→R3; its cost is €2750. We choose the first subtour for branching. Since it contains two arcs, C→R1 and R1→C, we generate two subproblems, P_1 in which we stipulate that arc C→R1 should be absent, and P_2 in which we stipulate that arc R1→C should not be present. At the end of this iteration, LIST = $\{P_1, P_2\}$, BEST = $\emptyset$, and BESTCOST = ∞. For this problem, the branch and bound method described here requires eight iterations. The details of these iterations are given in Table 6.3.

Since the traveling salesman problem is hard to solve in general, several heuristics have been developed for it. In the following, we describe a few of the heuristics which are commonly used to obtain good quality solutions to the traveling salesman problem within reasonable time. Since the network for the problem described in Section 6.1 is dense, we shall use the network in Figure 6.14 to illustrate the descriptions of the heuristics. Note that all the heuristics described here can get stuck on certain type of networks unless we consider edges that are absent in the network to be edges that are present but have infinite costs.

Nearest Neighbor

The *nearest neighbor heuristic* starts from any node in the graph, and chooses the node nearest to it. These two nodes are joined to form a partial tour. A third node that is closest to one of the two nodes is then chosen and joined to the node in the partial tour that is closest to it. This forms a partial tour with three nodes. Next a node that is closest to one of the endpoints of the partial tour is chosen, and joined to the partial tour. In this way the partial tour expands till there are no nodes left that are not in the partial tour. At this point, the end points of the partial tour are connected to each other to form a TSP tour. In some cases, this last connection is very expensive and gives rise to poor quality solutions.

Table 6.3. Branch and bound in action

Iteration	LIST	BEST BESTCOST (in €)	Sub-problem chosen	Bound (in €)	Decision
1	$\{P_0\}$	0	P_0	2750	Branch on C→R1→C
2	$\{P_1, P_2\}$	0	P_2	2825	Branch on R2→R5→R2
3	$\{P_2, P_3, P_4\}$	0	P_2	2825	Branch on R2→R5→R2
4	$\{P_3, P_4, P_5, P_6\}$	0	P_3	3250	Update BEST
5	$\{P_4, P_5, P_6\}$	3250 $T^\star$	P_4	3,927,992.49	Continue
6	$\{P_5, P_6\}$	3250 $T^\star$	P_5	4,071,420.00	Continue
7	$\{P_6\}$	3250 $T^\star$	P_6	4,185,000	Continue
8	$\emptyset$	3250 $T^\star$			Terminate

$T^\star = $ C→R4→R3→R5→R2→R1→C

P_0: Original problem   Assignment problem solution: {C→R1→C, R2→R5→R2, R3→R4→R3}

P_1: $\{P_0, y_{C,P_2} = 0\}$ Assignment problem solution: {R2→R5→R2, C→R3→R4→R1→C}

P_2: $\{P_0, y_{R4,C} = 0\}$ Assignment problem solution: {R2→R5→R2, C→R1→R4→R3→C}

P_3: $\{P_1, y_{R2,R5} = 0\}$ Assignment problem solution: {C→R4→R3→R5→R2→R1→C}

P_4: $\{P_1, y_{R5,R2} = 0\}$ Assignment problem solution: {C→R3→R4→R2→R5→R1→C}

P_5: $\{P_2, y_{R2,R5} = 0\}$ Assignment problem solution: {C→R1→R5→R2→R4→R3→C}

P_6: $\{P_2, y_{R5,R2} = 0\}$ Assignment problem solution: {C→R1→R4→R2→R5→R3→C}

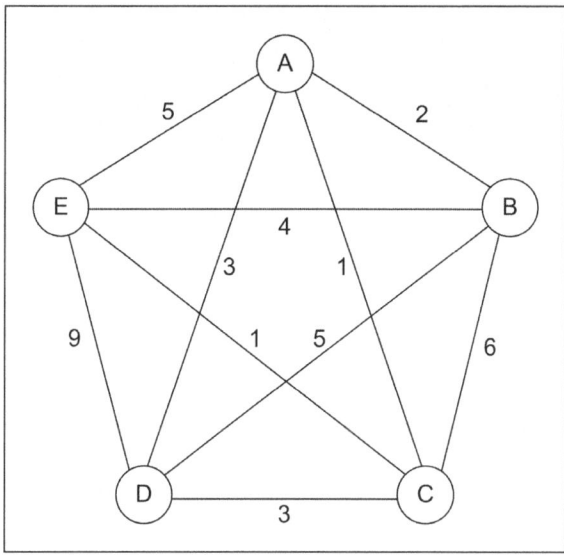

Fig. 6.14. Network used as example to explain heuristics

A description of the heuristic is given below, and the way it works for the graph in Figure 6.14 is shown in Figure 6.15. The thick lines refer to the current solution. The tour output by the nearest neighbor heuristic for the graph in Figure 6.14 is A – B – D – E – C – A with a length of 18 units.

Nearest Neighbor Heuristic

Input: A weighted graph $G = (V, E, d)$.

Output: A TSP tour T in G.

Step 1: Choose any node u in G, and find its closest neighbor v in the graph. Set $T \leftarrow (u, v)$, $end_1 \leftarrow u$, $end_2 \leftarrow v$, and $spanned \leftarrow \{u, v\}$. Go to Step 2.

Step 2: If there are no nodes in V that are not in $spanned$, then set $T \leftarrow T \cup (end_1, end_2)$, output T and terminate. Else go to Step 3.

Step 3: Choose a node $w \in V \setminus spanned$ such that it is a node in $V \setminus spanned$ that is closest to either end_1 or end_2. If w is closer to end_1 than end_2, then set $T \leftarrow T \cup (end_1, w)$, $spanned \leftarrow spanned \cup \{w\}$, and $end_1 \leftarrow w$. Else set $T \leftarrow T \cup (w, end_2)$, $spanned \leftarrow spanned \cup \{w\}$, and $end_2 \leftarrow w$. Go to Step 2.

Nearest Insertion

The *nearest insertion heuristic* is another popular heuristic for the TSP. At any iteration, a round trip involving a subset of all the nodes in the graph is maintained, and a new node added to the round trip. If any new node is to be added to the current trip, then one edge of the current trip needs to be deleted, and the two end points of the path thus formed need to be connected to the new node to re-create a round trip. The nearest insertion heuristic does this in the cheapest way possible at each

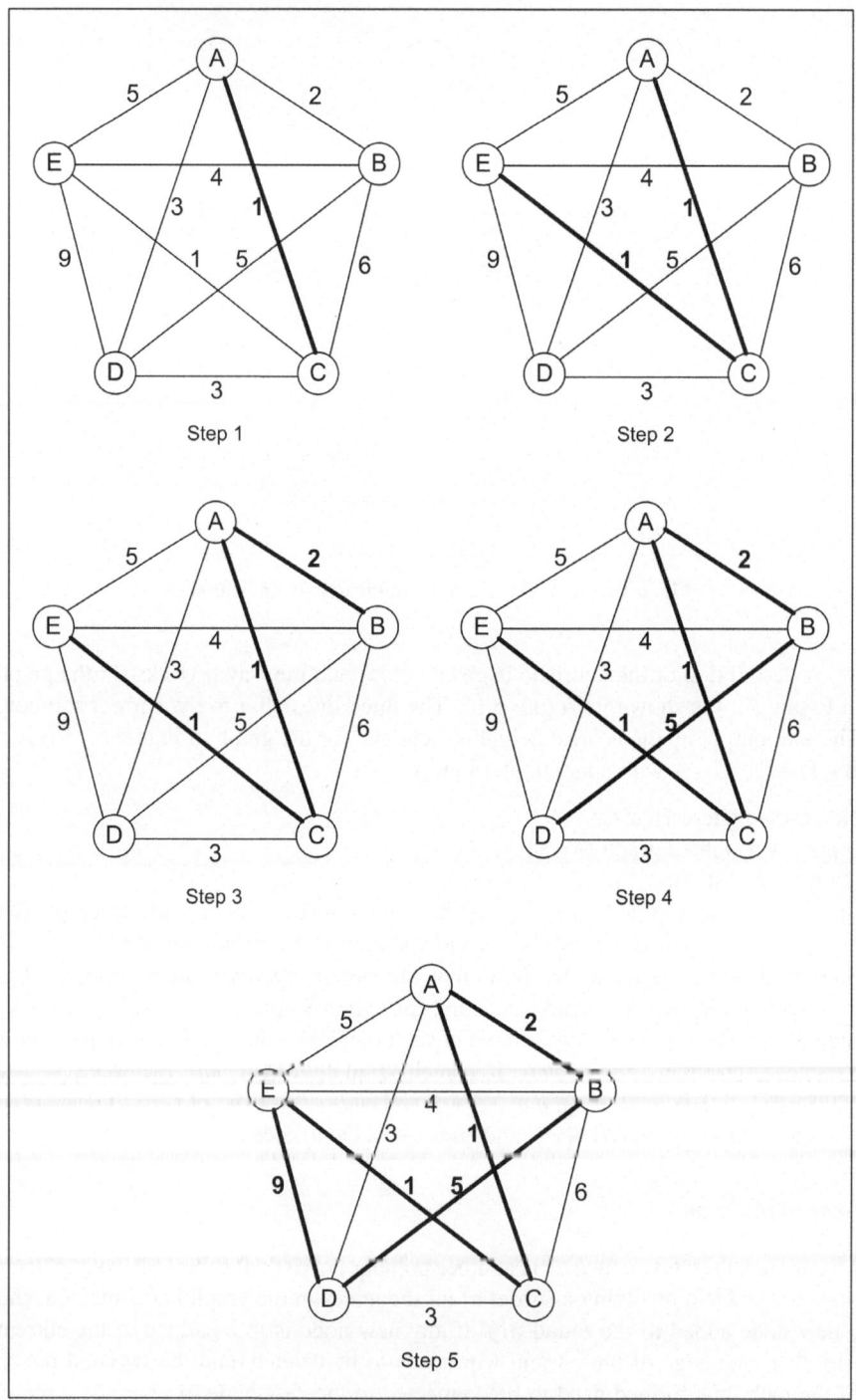

Fig. 6.15. Nearest neighbor heuristic in action

iteration. A description of the heuristic is given below, and the way it works for the graph in Figure 6.14 is shown in Figure 6.16. Here too, the thick lines refer to the current solution. The tour output by the nearest insertion heuristic for the graph in Figure 6.14 is A – B – E – C – D – A with a length of 13 units. Notice that this heuristic avoids the link D – E that contributed significantly to the cost of the tour output by the nearest neighbor heuristic.

Nearest Insertion Heuristic
Input: A weighted graph $G = (V, E, d)$.
Output: A TSP tour T in G.
Step 1: Choose any node u in G, and find its closest neighbor v in the graph. Set
$T \leftarrow \{(u,v), (v,u)\}$, and $spanned \leftarrow \{u,v\}$. Go to Step 2.
Step 2: If there are no nodes in V that are not in $spanned$, then output T and terminate. Else go to Step 3.
Step 3: For each node $w \in V \setminus spanned$ compute the cost of insertion of w, $cost_w = \min\{c(u,w) + c(w,v) - c(u,v) | (u,v) \in T\}$. Choose a node w^* with the minimum cost of insertion, and insert it in the tour as cheaply as possible. Set $spanned \leftarrow spanned \cup \{w\}$. Go to Step 2.

Farthest Insertion

The *farthest insertion heuristic* is a variant of the nearest insertion heuristic. At each iteration, it also starts with a round trip involving a subset of all the nodes in the graph, and a new node is added, but the new node is added in the most expensive way possible. A description of the heuristic is given below, and the way it works for the graph in Figure 6.14 is shown in Figure 6.17. Here too, the thick lines refer to the current solution. The tour output by the nearest insertion heuristic for the graph in Figure 6.14 is A – C – B – E – D – A with a length of 22 units. This is the worst of the three for this graph, but the farthest insertion heuristic returns solutions that are better, on average, than those returned by the nearest insertion heuristic!

Farthest Insertion Heuristic
Input: A weighted graph $G = (V, E, d)$.
Output: A TSP tour T in G.
Step 1: Choose any node u in G, and find its farthest neighbor v in the graph. Set
$T \leftarrow \{(u,v), (v,u)\}$, and $spanned \leftarrow \{u,v\}$. Go to Step 2.
Step 2: If there are no nodes in V that are not in $spanned$, then output T and terminate. Else go to Step 3.
Step 3: For each node $w \in V \setminus spanned$ compute the cost of insertion of w, $cost_w = \min\{c(u,w) + c(w,v) - c(u,v) | (u,v) \in T\}$. Choose a node w^* with the maximum cost of insertion, and insert it in the tour as cheaply as possible. Set $spanned \leftarrow spanned \cup \{w\}$. Go to Step 2.

The heuristics mentioned above are all *construction heuristics*, meaning that they have a TSP tour only at the end of their execution. There also are heuristics called *improvement heuristics*, that start out with a TSP tour, and try to improve the tour using

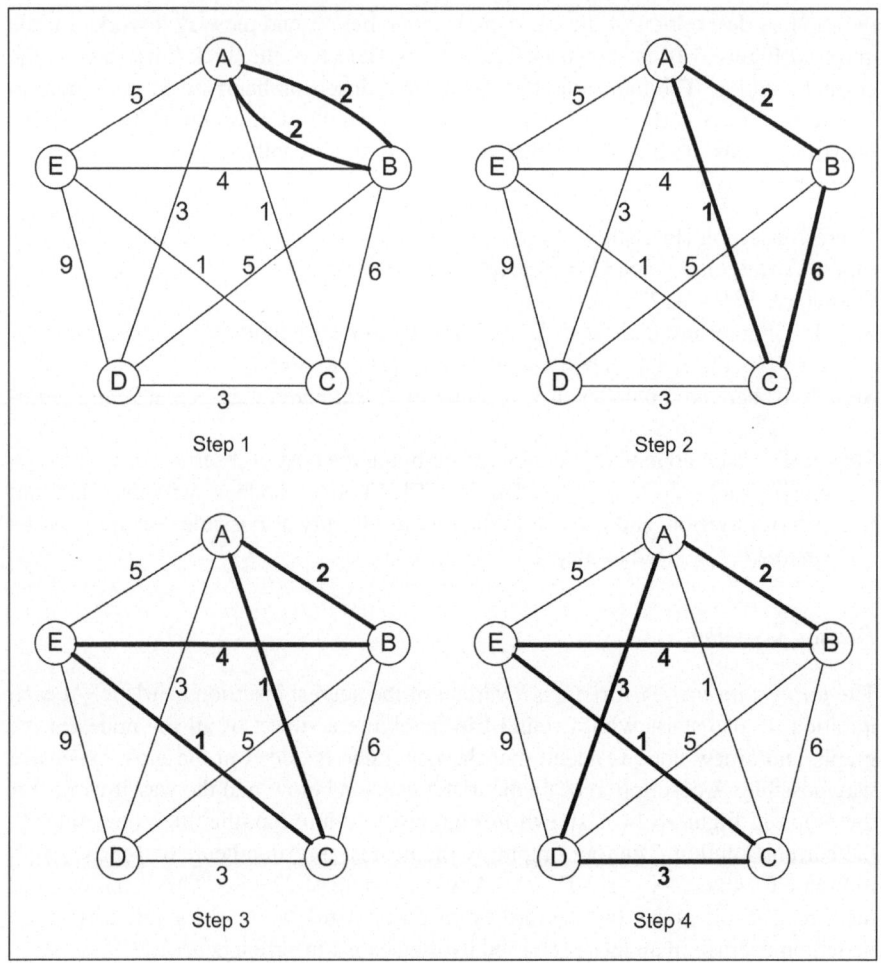

Fig. 6.16. Nearest insertion heuristic in action

certain fairly general methods. Here we describe one such improvement heuristic, namely the 2-opt heuristic.

2-Opt

The *2-opt heuristic* starts with a tour input by the user as the tour at hand, and searches its *neighborhood* for a better tour. If it finds a "better" tour (i.e., a tour with lower cost) in the neighborhood, then it calls this better tour the tour at hand, and again searches the neighborhood. The search stops when no tour in the neighborhood of the tour at hand is better than it. The neighborhood is defined by a set of *moves* which change the tour at hand. In the 2-opt heuristic, a move is defined as follows. Break the tour into two partial tours by removing two non-consecutive

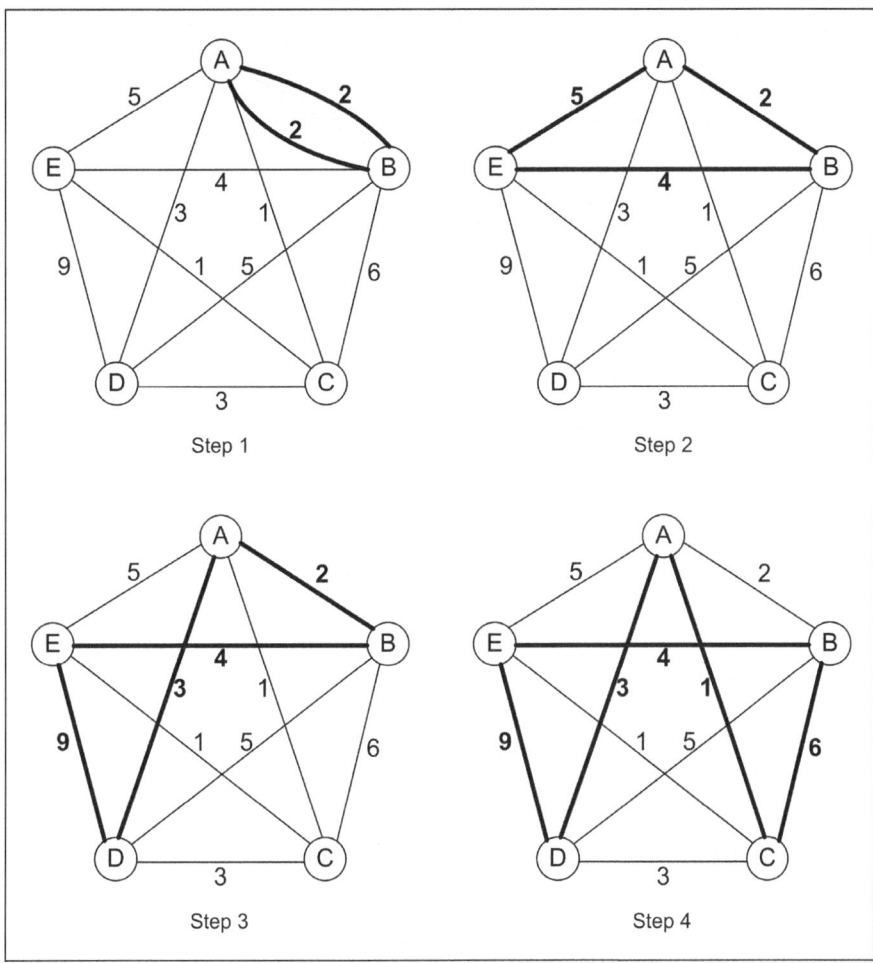

Fig. 6.17. Farthest insertion heuristic in action

edges, and then recombine the two partial tours using two different edges to form a neighboring tour. A 2-opt move is illustrated in Figure 6.18. The tour at hand is A – B – C – D – E – A. The partial tours E – A – B and C – D are formed by eliminating two edges (shown with dotted lines in the second diagram) and are reconnected to form the neighboring tour A – B – D – C – E – A. In this case, the original tour A – B – C – D – E – A had a cost of 25, while the neighboring tour A – B – D – C – E – A has a cost 16, which is better. This move therefore is called an *improving 2-opt move*. A description of the heuristic is given below, and the way it works for the graph in Figure 6.14 starting with the output of the farthest insertion heuristic as the input tour is shown in Figure 6.17. As in the previous cases, the thick lines represent the solution at hand. The 2-opt heuristic returns the tour A – B – E – C – D – A with a cost of 13 units.

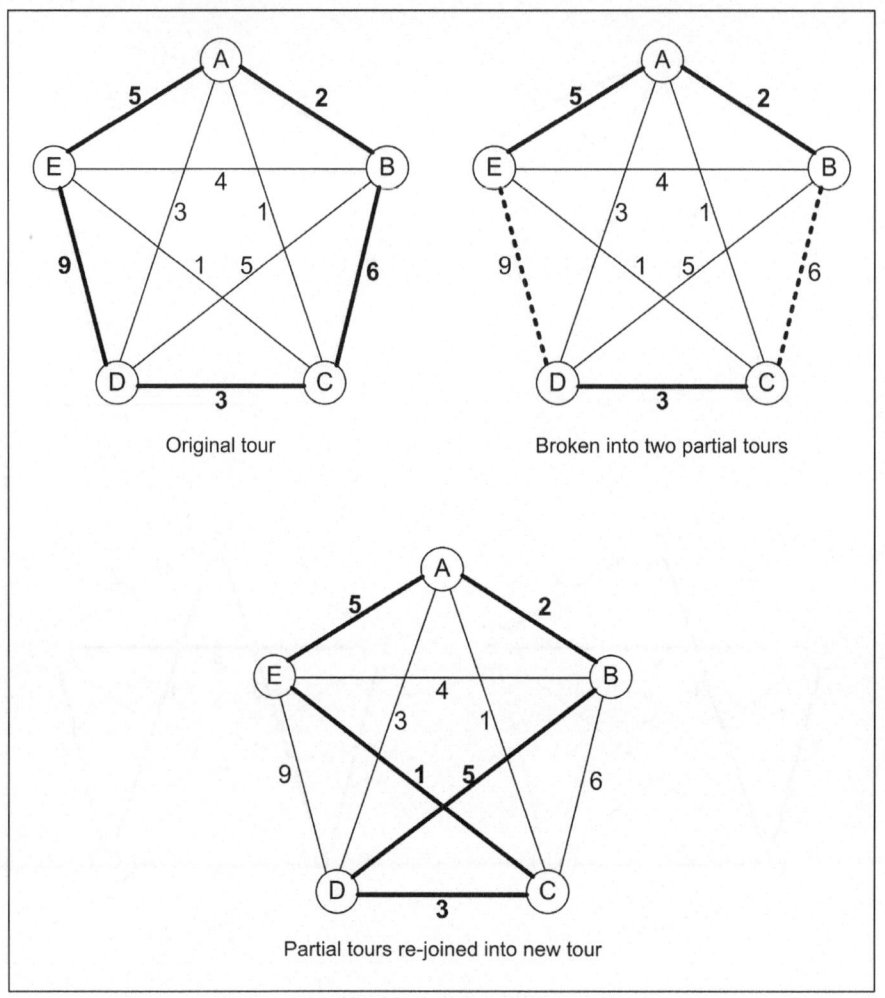

Fig. 6.18. A 2-opt move

2-Opt Heuristic

Input: A weighted graph $G = (V, E, d)$, and a TSP tour T_{in}.

Output: A 2-opt tour T in G.

Step 1: Set $T \leftarrow T_{in}$. Go to Step 2. (T is the tour at hand.)

Step 2: If no improving move is found from T, output T and terminate. Else go to Step 3.

Step 3: List all tours in the neighborhood of T that are obtained from T by improving 2-opt moves. Let T_{best} be the tour in the list with the lowest cost. Set $T \leftarrow T_{best}$. Go to Step 2.

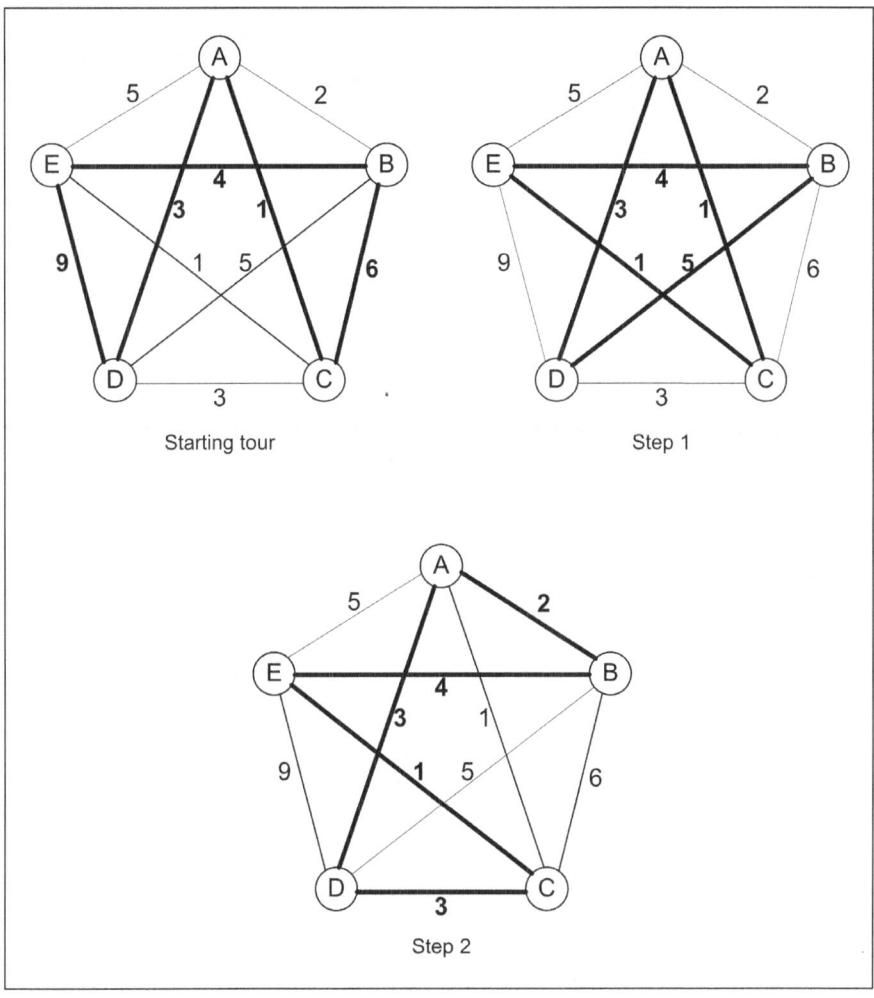

Fig. 6.19. 2-opt heuristic in action

6.4.2 Chinese postman problem

Contrary to the traveling salesman problem, the Chinese postman problem is not a hard problem except for mixed graphs, i.e., graphs containing both edges and arcs. This means that there are algorithms that solve all instances of the Chinese postman problem in time reasonable for the size of the problem.

Algorithms to solve the Chinese postman problem generally work in two phases. In the first phase, the network is augmented into an Eulerian network using the minimum number of copies of the edges in the network. In the second phase, the algorithm outputs a Chinese postman tour on the augmented network. Of course, if the original network is already Eulerian, the first phase is unnecessary.

The first phase

We start this phase with a network, say $N = (V, E)$, and augment it into an Eulerian network $N_E = (V, E_E)$ by adding the minimum number of copies of the edges in E. To do so, we first partition V into sets V_o consisting of nodes of odd degree in N, and V_e consisting of nodes of even degree in N. We next form a new complete weighted network N' on the nodes in V_o. To do this, for each pair of nodes i and j in V_o, we create an edge in N'. This edge corresponds to the path between i and j in N with the minimum number of edges in it, and the number of edges in this path is taken as the weight of the edge $i - j$ in N'. Once N' is constructed, we use a matching algorithm (see Chapter 4) to find a minimum weight matching in N'. This matching is a perfect matching, since the number of nodes in V_o is even, and the network N' is complete. We then create N_E as follows. First we copy each edge in E to E_E. Next, for each edge in N' that was part of the minimum weight matching, we add one copy of all the edges in N that were present in the path from i to j with the minimum number of edges. It can be seen that the network N_E is Eulerian. We next use N_E as an input for the second phase of the algorithm.

The second phase

In the second phase, we start with a network that is Eulerian, and find a tour on the network that traverses each of the edges in the network exactly once. We do this by constructing tours that progressively include all edges in the network. In the first iteration, we construct a tour on the network, not necessarily covering each of the edges in the network. Let us call this tour T_1. Now let the tour constructed at the end of iteration i be T_i. In the $(i+1)$th iteration, we choose a vertex v that is visited by the tour T_i, and also has an edge incident on it that is not a part of the tour. We construct an auxiliary tour T_a that starts from v and returns to it after traversing edges that are not part of T_i. Such a path exists because the network we are dealing with is Eulerian. We end the iteration by creating the tour T_{i+1} which combines T_i and T_a. To do this, we start T_{i+1} at vertex v, traverse the edges that made T_i to reach v, start out from v, this time traversing the edges that made T_a, and return to v again. We carry out these iterations until there is an iteration f such that T_f contains all the edges in the Eulerian network. The path thus formed is an optimal solution to the network we started the first phase with.

As an illustration, let us describe how this algorithm works on the network in Figure 6.4. In this network, V_o corresponds to $\{B, D\}$. The shortest path between B and D in the network is the direct arc B – D, and hence the Eulerian network formed at the end of the first stage is as shown in Figure 6.20. The second stage of the algorithm starts with the network in Figure 6.20. We see this network in the top left hand side of Figure 6.21. In the first iteration, we create the tour $T_1 = A - B - C - D$ in the network. This tour is shown in bold lines in the top right hand side of the figure. Next we start the second iteration. B – D is an edge in the network that is not part of T_1, but is incident on node B, which is visited by T_1. Thus we form the auxiliary tour B – D – B in the network which does not have any edge in common with T_1. This auxiliary tour is shown with broken lines in the diagram at the bottom

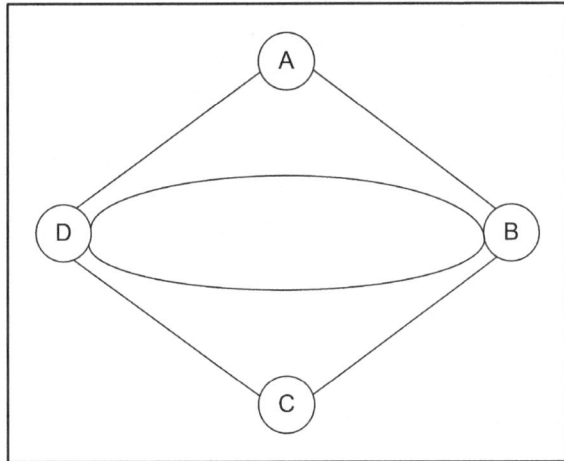

Fig. 6.20. Augmenting the network in Figure 6.4 to an Eulerian network

left side of Figure 6.21. Adding this to T_1, we obtain T_2, B – C – D – A – B – D – B. This tour is shown in the diagram at the bottom right side of Figure 6.21. Since this tour traverses all edges in the network, the algorithm stops. The optimal tour to the Chinese postman problem described by the network in Figure 6.4 is thus B – C – D – A – B – D – B.

6.5 Other Routing Problems

6.5.1 The bottleneck traveling salesman problem

Consider a sales representative in a company, who needs to meet clients at a number of locations connected to each other through a road network. Assume that the locations are in a sparsely populated region, so that, if the sales representative's car breaks down en route from one location to another, the sales representative would have a hard time getting the car repaired. In this situation, the sales representative would like to compute a round trip from his base location connecting all other locations and coming back to his original location in such a way that no leg of the trip is too large. The problem of finding such a round trip is known as the bottleneck traveling salesman problem.

Formally stated, in a *bottleneck traveling salesman problem* one is given a directed weighted graph $G = (V, A, W)$. The objective of the problem is to construct a simple cycle in G, passing through each node in V, such that the maximum of the weights of the arcs in the cycle is as small as possible.

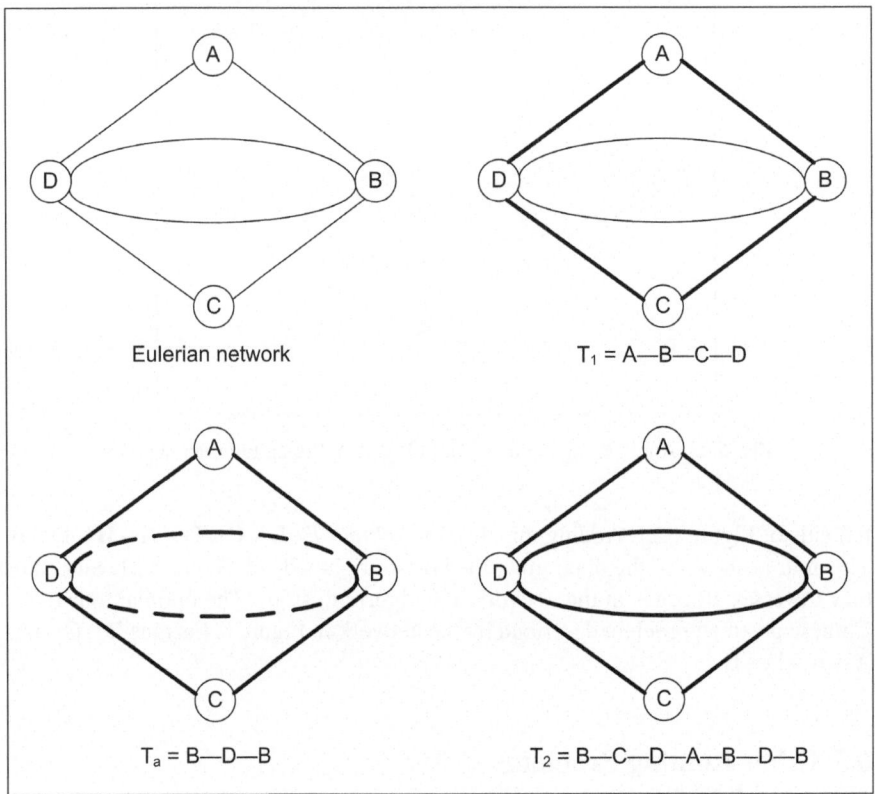

Fig. 6.21. Algorithm for the Chinese postman problem in action

6.5.2 The prize-collecting traveling salesman problem

Consider a factory that needs a certain quantity of raw material for its functioning. This raw material is available at multiple supply locations. Under contract, if the factory decides to collect material from a particular source, it has to collect the full stock of that source. The factory manager wants to find out the sequence in which raw material needs to be collected from the various suppliers to satisfy the requirements of the factory at minimum cost. This problem is called the prize-collecting traveling salesman problem. Note that in this problem all sources need not be visited by the factory.

Formally stated, in the *prize collecting traveling salesman problem* one is given a weighted directed graph $G = (V, A, W)$, in which each node $v \in V$ is assigned a non-negative weight p_v. The objective is to obtain a simple cycle C in N such that the sum of the weights at the nodes included in C exceeds a pre-specified number, and the sum of the weights on the arcs included in C is the minimum possible.

6.5.3 The generalized traveling salesman problem

Consider a multinational organization that has offices in several countries. Each country has at least one office, and some countries have multiple offices in different large cities. The organization's top management decides to hold meetings in each country for the executives in offices in that country. The cost of transporting the top management from one country to another is much more than the cost of assembling all the office managers of a country to a single city in the country. So the organization's problem is to choose a city in each country, and decide on a traveling salesman tour defined on all the chosen cities. This problem is called the generalized traveling salesman problem.

Formally stated, in the *generalized traveling salesman problem* one is given a weighted directed graph $G = (V, A, W)$, in which V is partitioned into k non-empty clusters. The objective is to output a simple cycle in G which visits each cluster at least once. In this context, a cycle visiting a cluster means that the cycle has one arc whose head is a particular node in the cluster and whose tail is a node outside the cluster, and another arc whose tail is that node in the cluster and whose head is outside the cluster. A common variant of the generalized traveling salesman problem is the *equality constrained generalized traveling salesman problem* in which a solution must visit each cluster exactly once.

6.5.4 The rural postman problem

In a garbage disposal system, the streets in a city are partitioned into seven sets. Garbage is collected from the houses along the streets in a particular set on one pre-specified day of the week. The manager of the garbage disposal system needs to find an efficient way to route garbage trucks on each day of the week so that the collection of garbage requires minimum time. This problem can be modeled as a rural postman problem.

In a *rural postman problem*, one is given a weighted graph $G = (V, E, W)$, and a set $R \subset E$ of edges of the graph. A solution to the problem is a route through G covering each of the edges in R. The route can pass through the same edge more than once, if required. The cost of a solution is the product of the weight of an edge and the number of times the route passes through it, summed over all the edges in the graph. The objective is to find a minimum cost solution.

6.6 Exercises on Network Routing Problems

Problem 6.1. Production Scheduling

Global Handset Company (GHC) is a subsidiary of GTC that manufactures handsets for GTC customers. These handsets come in various colors, and GHC works as a job shop, accepting orders from GTC and delivering them within a desired date. GHC works 8 hours a day, 5 days a week, and uses injection molding technology to manufacture the plastic components in the handsets and are capable of manufacturing 500 handsets each hour. The assembly of the handsets requires very little time, so that the time required to produce a handset depends critically on the number of handset components that are molded. GHC requires time to change the colors of handsets, and these changeover times (in minutes) are given in Table 6.4. On Friday evening,

Table 6.4. Changeover times

	Clear	White	Yellow	Orange	Red	Purple	Blue	Green	Black
Clear	—	15	15	15	15	15	15	15	15
White	49	—	32	50	49	49	50	50	49
Yellow	50	47	—	18	35	47	63	62	65
Orange	65	50	48	—	48	53	64	80	64
Red	63	78	48	34	—	18	79	79	65
Purple	79	80	65	65	47	—	80	79	50
Blue	63	64	77	78	65	65	—	19	49
Green	62	63	77	77	65	64	32	—	47
Black	93	92	92	94	77	63	93	92	—

GHC received an order for 18000 handsets to be delivered next Friday (5 working days later). The details of the order are given in Table 6.5. GHC knows that it needs

Table 6.5. Work order for GHC

Clear	White	Yellow	Orange	Red	Purple	Blue	Green	Black
6000	4000	1000	3000	1000	500	500	500	1500

to pay overtime to be able to meet the order. However, by sequencing the production activity properly, it can reduce the amount of overtime it must pay. (GHC pays overtime at the rate of €500 per hour)

(a) Formulate this problem as a weighted Hamiltonian path problem.
(b) Obtain a feasible solution to the problem using the nearest neighbor heuristic starting with the production of handsets colored "clear".
(c) Compute a production plan for GHC that requires the lowest production and changeover time.

At the beginning of day 3 of production, GTC informs GHC that there has been a mistake, and that the demand for "clear" handsets should be 7000 instead of 6000. Correspondingly, the demand for "black" handsets should be 500 instead of 1500.

(d) Will this change the production plan? If so, what are the cost implications of this mistake?

Problem 6.2. Audit Team Movement

A large scale audit has been proposed for the 10 units of GTC (named A, B, ..., J). These are located in geographically diverse regions of the country, and the cost of transporting (in €s) the audit team from one unit to the other is given in Table 6.6. The audit team starts from the base unit at G.

Table 6.6. Transportation costs between units

	A	B	C	D	E	F	G	H	I	J
A	—	200	600	447	632	283	283	447	566	400
B		—	632	632	447	200	447	283	447	447
C			—	894	721	447	447	894	447	1000
D				—	1077	721	447	800	1000	447
E					—	400	800	447	283	849
F						—	400	447	283	632
G							—	721	632	632
H								—	600	447
I									—	894
J										—

(a) What would be the cheapest way of organizing the audit process if the team has to return to the base unit at G?

(b) The team would like to visit units A, B, G, and J before they visit any of the other units. Under this consideration, what would be the cheapest way of organizing the audit process?

(c) The audit team is told that the unit of A is a subsidiary of the unit at G, and the unit at F is a subsidiary of the unit at B. The audit team therefore would like to visit A immediately after visiting G, and F immediately after visiting B. Under this consideration, and without the consideration in (b) above, what is the cheapest way of organizing the audit?

Problem 6.3. Scheduling Discussion Meetings

The maintenance department of GTC is executing 14 projects at the moment. These projects are labeled P1, ..., P14. The supervisor of the department, Ms. Thomas, has chosen to meet the seven employees that work on the projects. The employees are labeled A1, ..., A7. An employee may be involved in more than one project. Table 6.7 lists the employee project relations. From this table, we can see that, for example, employee A5 is involved in projects P7, P10, P12, and P14. In order to schedule the meetings most efficiently, Ms. Thomas has to ensure that the total number of times that employees have to go and come back for discussing another project has to be minimized. For instance, for employee A2, it is most efficient if projects P1, P2, P3, P7, and P9 are discussed consecutively. In other words, Ms. Thomas wants to reduce traffic in and out of her office.

Table 6.7. Employee involvement in projects

Projects	Employee						
	A1	A2	A3	A4	A5	A6	A7
P1	1	1					
P2	1	1					1
P3		1	1				
P4	1		1	1			
P5			1	1			
P6			1				
P7	1			1			
P8						1	1
P9	1					1	1
P10			1	1			
P11			1				
P12			1	1	1		
P13	1	1					
P14						1	1

(a) Formulate Ms. Thomas' scheduling problem as a TSP. What is the "distance matrix" in this case.
(b) Use the nearest neighbor heuristic to find a feasible meeting schedule, starting with project P1.
(c) Try to improve the solution in part (b) by means of the 2-opt heuristic.
(d) Apply the farthest and nearest insertion heuristic for finding feasible schedules. Try to improve these solutions by means of the 2-opt heuristic. What is your best solution found so far?
(e) Employee A4 has been recently released from project P5. What happens to the distance matrix you constructed in part (a)?
(f) Construct a schedule of meetings to reduce traffic in and out of Ms. Thomas' office using the nearest insertion heuristic for the case when employee A4 has been released from project P5. Apply the 2-opt heuristic to the output of the nearest neighbor heuristic.
(g) Try to find optimal solutions to the scheduling problem with distance matrices developed in parts (a) and (c).
(h) Try to construct a meeting scheduling problem in which the distance matrix has all non-zero entries, and all entries have the same value. How many optimal schedules does such a problem have?

Problem 6.4. Traveling in Music Land

GTC is planning to sponsor the development of a theme park, named Music Land.

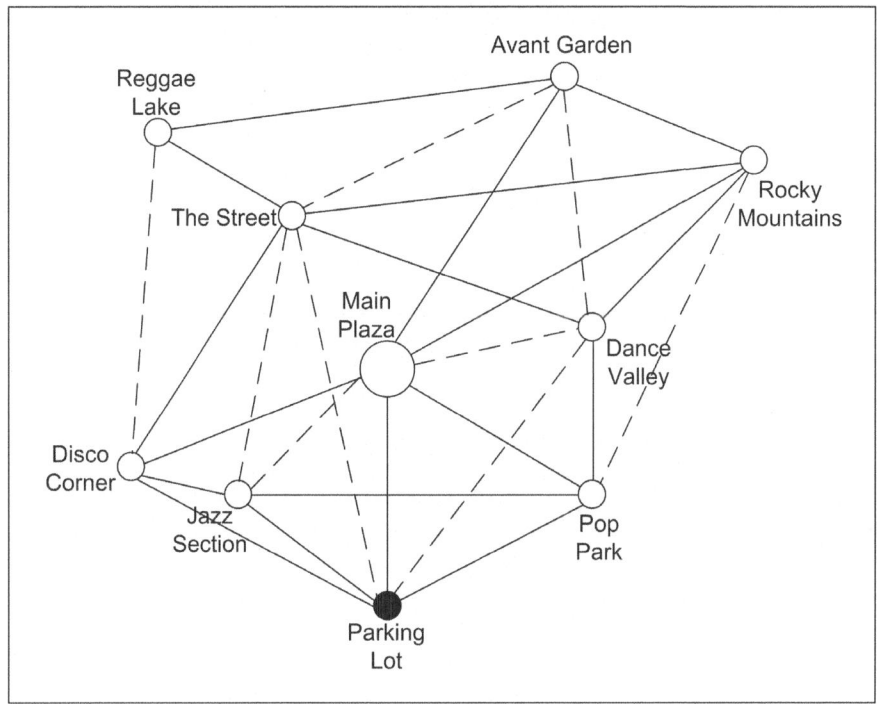

Fig. 6.22. Map of Music Land

The idea is to build theme areas based on specific music styles, called Jazz Section (JS), Disco Corner (DC), Reggae Lake (RL), The Street (TS), Avant Garden (AG), Rocky Mountains (RM), Dance Valley (DV), and Pop Park (PP). Each of these areas will have a theme restaurant, amusement attractions, and souvenir shops. In addition, the park will include a Main Plaza (MP) with a multimusical atmosphere, and a Parking Lot (PL). A map of the theme park is shown in Figure 6.22. The solid lines between the areas denote pathways that can be used both by pedestrians and by motorized transport. The dashed lines denote pathways that can be used only by motorized transport. At points where paths cross in Figure 6.22 (for example, the path between TS and DV 'crosses' the path between MP and AG) there are tunnels which ensure that the traffic flows do not interfere. The lengths of the various pathways (in meters) are given in Table 6.8. Since many of the visitors at Music Land prefer to walk through the park, GTC would like to offer each visitor a brochure with specific walking tours. The start and finish of all tours should be the Parking Lot, and they should include all theme areas and the Main Plaza.

Table 6.8. Length of pathways (in meters)

	PL	JS	MP	DC	RL	TS	AG	RM	DV	PP
PL	—	1050	1200	1635	—	1650	—	—	1725	840
JS		—	555	840	—	645	—	—	—	1320
MP			—	1365	—	585	1230	1875	975	1035
DC				—	750	1095	—	—	—	—
RL					—	855	1695	—	—	—
TS						—	1095	2070	1440	—
AG							—	1425	1260	—
RM								—	795	1740
DV									—	1005
PP										—

(a) What is a shortest walking tour covering all theme parks, the Main Plaza, and the Parking Lot?

GTC is also thinking of suggesting a scenic walking tour for visitors to Music Land. It has already collected data from 100 volunteers, who were asked to rate each pedestrian pathway with a score from 1 through 10, in which 10 denotes the highest rating (most scenic). Table 6.9 shows the average of these ratings rounded to the nearest integer.

Table 6.9. Volunteer valuation of pedestrian pathways

	PL	JS	MP	DC	RL	TS	AG	RM	DV	PP
PL	—	6	7	6	—	—	—	—	—	6
JS	6	—	—	5	—	—	—	—	—	8
MP	7	—	—	6	—	7	8	7	—	8
DC	6	5	6	—	—	9	—	—	—	—
RL	—	—	—	—	—	9	8	—	—	—
TS	—	—	7	9	9	—	—	8	7	—
AG	—	—	8	—	8	—	—	7	—	—
RM	—	—	7	—	—	8	7	—	7	—
DV	—	—	—	—	—	7	—	7	—	8
PP	6	8	8	—	—	—	—	—	8	—

(b) What is a most scenic walking tour covering all theme parks, the Main Plaza, and the Parking Lot?
(c) GTC feels that some visitors would want a walking tour covering all pathways meant for pedestrians. What would be the length of a shortest tour covering all pathways that allow pedestrians? Which pathways would be traversed more than once?

GTC also is thinking about having two motorized modes of transportation in Music Land: streetcar and monorail. These systems can be built only in regions suitable for motorized transport. However, the cost of building these modes depend on the nature and gradient of the region. In case the area is rocky, then the cost of laying tracks is a certain percentage more than if the region was normal. This extra percentage is called "rocky area bonus". If the region has a gradient, then there is no extra cost involved for laying the tracks, but there is a stipulated maximum gradient, that each of the motorized modes can handle.

Most of the area in Music Land can be called normal. However, the pathways between Jazz Section and Main Plaza, and between Avant Garden and The Street are built on rocky areas. The gradient of the pathway from Rocky Mountains to Dance Valley is 7.1% up. In addition, a bridge needs to be built between Reggae Lake and Disco Corner at a cost of €1,000,000, if that pathway has to be used by any of the motorized modes planned.

GTC has been negotiating with PSC Streetcars for the streetcar system. PSC Streetcars can handle gradients of 8% both up and down. PSC has suggested the cost structure in Table 6.10 for its streetcar system.

Table 6.10. PSC Streetcar cost structure

Normal capital cost*	€950,000/km
Rocky area bonus	10%
Cost of one streetcar	€50,000

$\star$: This includes the cost of building stations.

For the monorail system, GTC has been negotiating with two firms, MNRA and EMCL. Table 6.11 summarizes the relevant information provided by the two firms in the course of their negotiations with GTC.

Table 6.11. Information provided by MNRA and EMCL

	MNRA	EMCL
Normal capital cost*	€13,000,000/km	€15,000,000/km
Rocky area bonus	15%	10%
Station building costs	€7,200,000/station	€4,300,000/station
Stipulated max. grad.	12%	4.6% up, 6.6% down
Number of trains	2 (one in each direction)	1
Average speed	33 km/hr	28 km/hr

$\star$: This includes the cost of buying the train(s) but excludes cost of building stations.

(d) Suggest a design for the motorized transport modes. Which monorail company is more profitable for Music Land?

(e) There is a possibility of laying a new road from Rocky Mountains to Dance Valley at a cost of €1,500,000. This road will have a gradient of 5% up, and will pass through rocky areas. The length of the road will be 900 meters. Does it make sense to construct this road before the motorized transport modes are built?

Problem 6.5. Designing Bus Routes

GTC has a research wing set up in the city marked A in Figure 6.23. The personnel for this wing come from the neighboring villages, marked from B through P in the same figure. The figure also schematically provides the road network between these villages, with the length of the road segment between a pair of villages (in kilometers) marked on the line joining the two. For example, villages B and C are connected by a road 1.0 kilometer long.

GTC wants to make use of a bus service to pick up all the personnel from their villages at the start of the day, and drop them back after work. It is negotiating a contract with a local bus service company BSC. The bus service company has given GTC two options.

In the first option, BSC provides GTC with a long distance bus, which starts at A and with which GTC can pick up all its personnel in one round trip and deposit them to its research wing at A. The cost of hiring the bus is €2000 per month, and BSC will charge €7.50 per kilometer traveled by the bus in each month.

In the second option, BSC will provide GTC with two minibuses. These minibuses can travel up to 10 kilometers in one trip. Using these buses, GTC has to plan two round trips to transport all its personnel to and from the research wing. The cost of hiring each minibus is €1050 per month and BSC will charge as usual €7.00 per kilometer traveled by the bus in each month.

On an average, GTC works 20 days in a month.

(a) GTC wants to figure out the approximate cost per month of operating under the first option. Starting at city A, use the nearest insertion heuristic to come up with a feasible route and the cost of operating this route. What do you observe?
(b) Find the lowest cost that GTC would have to pay to operate under the first option.
(c) Find by inspection, two routes that the GTC could operate under if they chose the second option. What is the rationale behind your choice? What is the cost of operating these two routes?
(d) Using these two routes as a starting solution, and transforming the problem into a TSP, see if you can apply the 2-opt heuristic to improve the routes obtained in part (c).
(e) What is the lowest cost that GTC would have to pay to operate under the second option. How different is it from your answer in part (d)? Based on your result should GTC use the first option or the second one?
(f) GTC argues that the requirement of round trips is constraining the routes chosen under the second option Can you provide an example in support of GTC's argument?

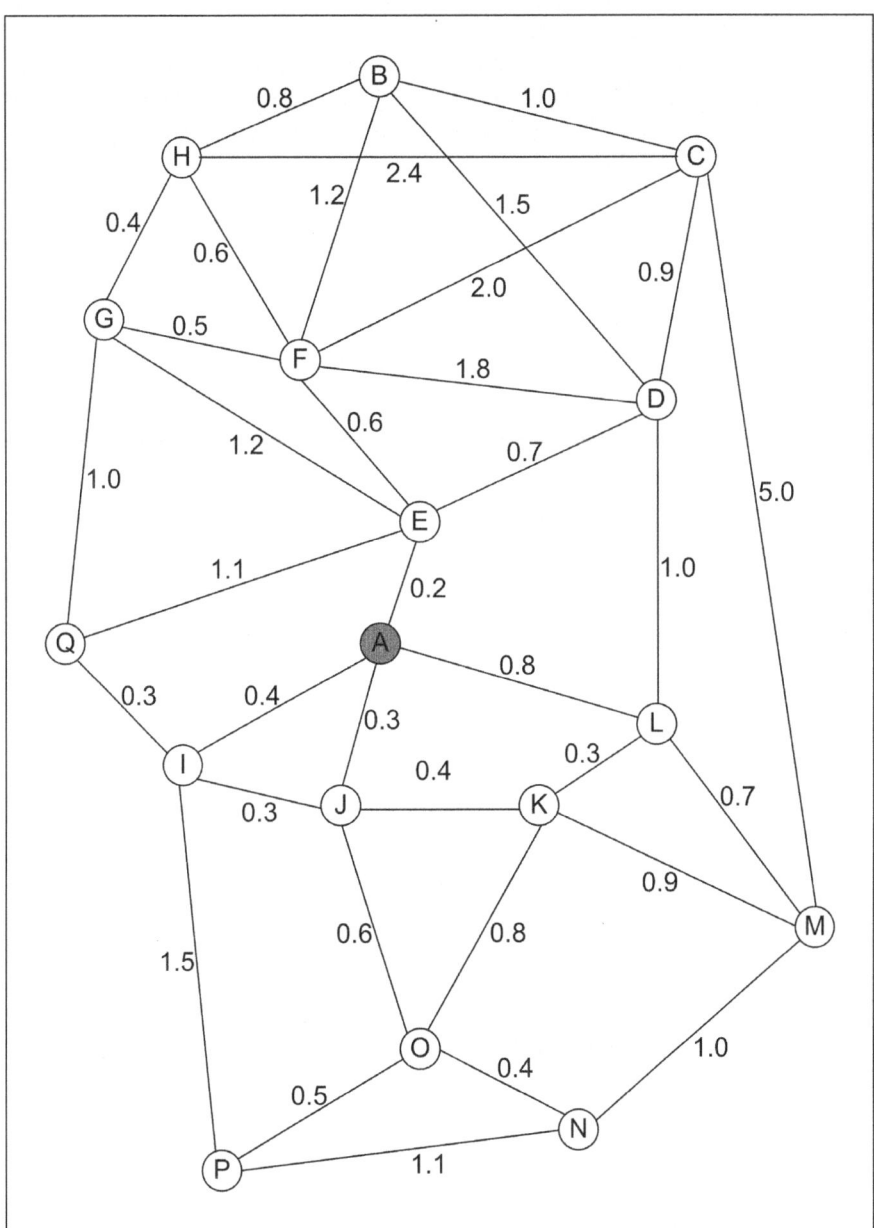

Fig. 6.23. Distance map for the region

Problem 6.6. Washing Roads

GTC has set up operations in a developing country. As an advertising measure, they have decided to take up the responsibility of washing the road network in the downtown area of the capital of the country, and charge the municipality only for the fuel costs they incur for the operation. This area consists of nine points (A, B, ..., I) and a road network connecting them. GTC has estimated the fuel costs for traversing each of these roads, and the costs are given (in € cents per day) in Figure 6.24.

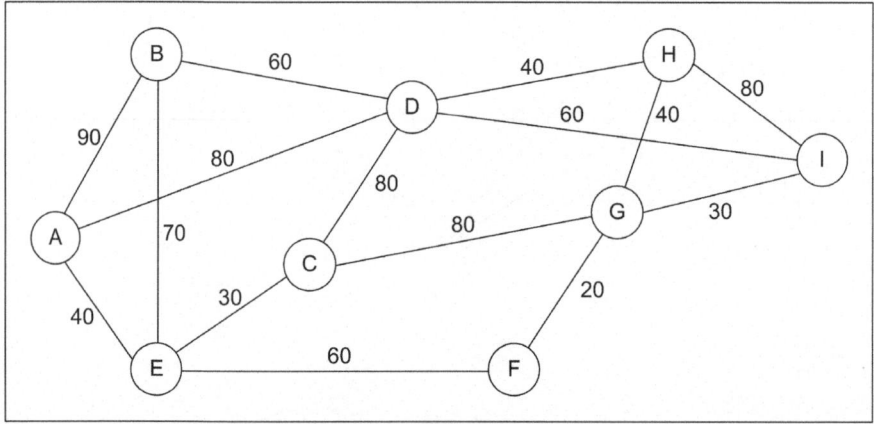

Fig. 6.24. Road network and fuel costs

(a) Since this cost is to be incurred every day for a long time, GTC wants to compute the most efficient route to employ in cleaning the area. What is the minimum per day that GTC should be prepared to incur in fuel costs?

(b) There is some repair work being done on segments between A and D and between G and H. During the time this repair work is on, GTC does not need to wash these segments. The municipality wants to know if GTC would reduce their daily charge, and if so, by how much?

(c) Consider the situation in part (a). The municipality informs GTC that they are planning a new road between points B and H, and wants GTC to take the responsibility of cleaning this segment at no extra cost. GTC estimates that the fuel cost for traversing this road segment is €0.30 per day. Should GTC agree on this extension of contract?

(d) If the municipality finished constructing the road between B and H while the repair work was being done on segments between A and D and between G and H. If GTC is to take up the road washing activity on the road network including the new road but excluding the segments on which repair work was being done, how much fuel costs would GTC incur per day?

References

1. E. Aarts, J.K. Lenstra (Eds.) (1997), *Local Search in Combinatorial Optimization*, John Wiley & Sons, New York.
2. R.K. Ahuja, T.L. Magnanti, J.B. Orlin (1993), *Network Flows; Theory, Algorithms and Applications*, Prentice Hall.
3. J.M. Aldous, R.J. Wilson (2001), *Graphs and Applications; an Introductory Approach*, Springer.
4. M.S. Bazaraa, J.J. Jarvis, H.D. Sherali (1990), *Linear Programming and Network Flows*, 2nd Edition, John Wiley & Sons, New York.
5. L.D. Bodin, B.L. Golden, A.A. Assad, M.O. Ball (1983), Routing and Scheduling of Vehicles and Crews; the State of the Art, *Computers and Operations Research* 10, pp. 69-211.
6. J.A. Bondy, U.S.R. Murty (1977), *Graph Theory with Applications*, American Elsevier Publishing Co., Inc.
7. B.H. Boon, G. Sierksma (2003), Team Formation: Matching Quality Supply and Quality Demand, *European Journal of Operational Research* 148, pp. 277-292.
8. D.J. Bowersox, D.J. Closs (1996), *Logistical Management: the Integrated Supply Chain Process*, The McGraw-Hill Companies, Inc.
9. J. Bramel, D. Simchi-Levi (1997), *The Logic of Logistics; Theory, Algorithms, and Applications for Logistics Management*, Springer-Verlag, New York.
10. W.J. Cook, W.H. Cunningham, W.R. Pulleyblank, A. Schrijver (1998), *Combinatorial Optimization*, John Wiley & Sons, Inc.
11. V. Chvátal (1983), *Linear Programming*, W.H. Freeman and Company, New York.
12. M.S. Daskin (1995), *Network and Discrete Location; Models, Algorithms, and Applications*, John Wiley & Sons, Inc.
13. D.-Z. Du, P.M. Pardalos (1998) *Handbook of Combinatorial Optimization, Volumes 1, 2, and 3*, Springer.
14. J.R. Evans, E. Minieka (1992), *Optimization Algorithms for Networks and Graphs*, 2nd Ed., Marcel Dekker, Inc., New York.
15. G. Gutin, A.P. Punnen (Eds.) (2002), *The Traveling Salesman Problem and its Variations*, Kluwer Academic Publishers.
16. F.S. Hillier, G.J. Lieberman (1995), *Introduction to Operations Research*, 6th Ed., McGraw-Hill, Inc.
17. P.A. Jensen, J.F. Bard (2003), *Operations Research; Models and Methods*, John Wiley & Sons, Inc.
18. D. Jungnickel (1999), *Graphs, Networks, and Algorithms*, Springer-Verlag, Berlin.
19. B. Korte, J. Vygen (2002), *Combinatorial Optimization; Theory and Algorithms*, 2nd Ed., Springer-Verlag.
20. E.L. Lawler, J.K. Lenstra, A.H.G. Rinnooy Kan, D.B. Shmoys (Eds.) (1990), *The Traveling Salesman Problem; a Guided Tour of Combinatorial Optimization*, John Wiley & Sons, Chichester.
21. J. Lee (2004), *A First Course in Combinatorial Optimization*, Cambridge University Press.
22. P.B. Mirchandani, R.L. Francis (1990), *Discrete Location Theory*, John Wiley & Sons, Inc.
23. J.M. Padberg (1995), *Linear Optimization and Extensions*, Springer-Verlag.
24. C.H. Papadimitriou, K. Steiglitz (1982), *Combinatorial Optimization: Algorithms and Complexity*, Prentice-Hall, Inc., Engelwood Cliffs, N.J.

25. A. Schrijver (1986), *Theory of Linear and Integer Programming*, John Wiley & Sons, Chichester.
26. A. Schrijver (2003), *Combinatorial Optimization: Polyhedra and Efficiency, Volumes A, B, and C*, Springer.
27. G. Sierksma (2002), *Linear and Integer Programming; Theory and Practice*, 2nd Ed., Marcel Dekker, Inc., New York.
28. G. Sierksma, G.A. Tijssen (1998), Routing Helicopters for Crew Exchanges on Offshore Locations, Mathematics of Industrial Sciences, Vol. 3, *Annals of Operations Research*, Baltzer's Publishers.
29. D. Simchi-Levi, P. Kaminski, E. Simchy-Levi (2003), *Designing and Manging the Supply Chain; Concepts, Strategies, and Case Studies*, McGraw-Hill.
30. H.A. Taha (2003), *Operations Research: an Introduction*, 7th Ed., Pearson Education, Inc., Upper Saddle River, N.J.
31. H.P. Williams (1990), *Model Building in Mathematical Programming*, John Wiley & Sons, Chichester.
32. H.P. Williams (1993), *Model Solving in Mathematical Programming*, John Wiley & Sons, Chichester.
33. R.J. Wilson, J.J. Watkins (1990), *Graphs: an Introductory Approach*, John Wiley & Sons, Inc.
34. W.L. Winston (2003), *Operations Research: Applications and Algorithms*, 4th Ed., Thomson Brooks/Cole, Thomson Learning, Inc.
35. L.A. Wolsey (1998), *Integer Programming*, John Wiley & Sons, Inc.

Index